Waste-to-Energy

Waste-to-Energy
Technologies and Project Implementation

Second Edition

Marc J. Rogoff and Francois Screve

AMSTERDAM • BOSTON • HEIDELBERG • LONDON • NEW YORK
OXFORD • PARIS • SAN DIEGO • SAN FRANCISCO • SINGAPORE
SYDNEY • TOKYO

William Andrew is an imprint of Elsevier

William Andrew is an imprint of Elsevier
225 Wyman Drive, Waltham, MA 02451, USA
The Boulevard, Langford Lane, Kidlington, Oxford, OX5 1GB, UK

First edition 1987 by Noyes Publication, USA
Second edition 2011

Library of Congress Cataloging in Publication Data
A catalog record for this book is available from the Library of Congress

British Library Cataloguing in Publication Data
A catalogue record for this book is available from the British Library

ISBN: 978-1-4377-7871-7

For information on all Elsevier publications visit our website at www/elsevierdirect.com

Typeset by TNQ Books and Journals Pvt. Ltd.

Transferred to Digital Printing in 2011

Contents

Preface

A waste-to-energy (WTE) facility is perhaps the single most complex public works project usually attempted by a community. For successful implementation, it needs the services of knowledgeable in-house staff and consultants and active support of the community's decision-makers and public.

Why then should communities undertake such projects in the wake of such potential obstacles? The answer lies in the fact that many communities across this globe have no other answer to their problems of mounting increases in waste volumes, difficulty in siting alternative waste disposal facilities, and an ever-increasing need for renewable energy supplies. With more than 1,300 such facilities in operation around the world today with an installed capacity of 505,000 tpd or 160 million tonnes of waste turned into energy each year, WTE is an increasingly viable alternative for the world's rapidly urbanizing communities. The appendix at the end of the book contains Case Studies of WTE facilities.

This book is a major revision of an earlier edition, which was published in 1987. Many changes have occurred in the WTE industry worldwide since that time, including changes to technologies available, improvements in air emission control devices, new emission regulations imposed by the US Environmental Protection Agency and the European Union, a new economical and financial environment and the movement worldwide to confront climate change.

During the development of this new edition, many industry experts assisted us in reviewing draft chapters and contributed valuable data and information. We would like to specifically acknowledge Bruce Clark, Fred Caillard, Charlie Tripp, and Ken Boatwright.

Finally, our hope is that this revised edition will provide sufficient information on what steps must be undertaken to implement a WTE project. While every project is unique to some degree, our hope is that the lessons learned over the past few decades with WTE project implementation will lead to more successful projects in the future.

Marc J. Rogoff
Tampa, Florida

Francois Screve
Ross, California

About the Authors

Marc J. Rogoff is a Project Director with **SCS Engineers**. He has over 30 years of experience in solid waste management as a public agency manager and consultant. He has managed more than 200 consulting assignments across the United States on literally all facets of solid waste management, including waste collection studies, facility feasibility assessments, facility site selection, property acquisition, environmental permitting, operation plan development, solid waste facility benchmarking, ordinance development, solid waste plans, financial assessments, rate studies/audits, development of construction procurement documents, bid and RFP evaluation, contract negotiation, and bond financings.

Dr. Rogoff has directed engineers' feasibility reports for nearly two dozen public works projects totaling more than $1.2 billion in project financings. He has interacted with bond rating agencies, financial advisors, insurance underwriters and investment bankers involved in these financings. His efforts have included the development of detailed spreadsheet rate models establishing the financial feasibility of each projects, long-term economic forecasts, and projected rate impact upon project users and customers.

Dr. Rogoff has extensive experience in the development of WTE projects from the initial feasibility to commercial operations monitoring. He has conducted bond feasibility studies and operations assessments, and provided recommendations on key procurement issues. Dr. Rogoff has conducted feasibility studies on more than 50 facilities worldwide.

Francois Screve holds 25 years of multi-national expertise in design and operation of solid waste fuel power plants in Asia, North America, and Europe. He is a results-oriented leader with a record of building high performance organizations. Mr. Screve has extensive experience in the development, optimization, and management of Energy-from-Waste and biomass projects from the initial feasibility to general design, commissioning, and commercial operations. His firm, **Deltaway Energy, Inc.**, has provided engineering and management services to more than 25 facilities in 13 countries worldwide since 2003. Before he started Deltaway Energy, Inc., Mr. Screve worked for 20 years for Veolia Environnement in Europe, the USA, and Asia, with the management of two large WTE facilities in California and Florida and the responsibility for all solid waste treatment operations for Veolia in Asia (11 countries from India to Japan) including the operation of seven WTE facilities and construction of three new facilities in Taiwan.

1 Introduction and overview

Chapter Outline

1.1 The growing solid waste disposal problem

Over the past several decades, various non-governmental organizations (NGOs) – such as the Asian Development Bank, United Nations, World Bank – and international research agencies supported by various European nations have estimated that solid waste that is generated worldwide may total between 2.5 to 4 billion metric tons [1]. However, on the basis of this extensive research on the growing solid waste problem, these investigators have also concluded that it is currently impossible to arrive at a more accurate estimate given the paucity of existing waste stream data in the developing nations, and the inaccuracy of common definitions of different waste streams from country to country. Inadequate collection and uncontrolled disposal of solid waste is extremely serious in low- (US$905 per capita or less) and middle-income (US$906 to US$11,115 per capita) countries, especially in rural areas, where solid waste is disposed of in uncontrolled dumps.

What appears somewhat more reliable are: estimates of municipal solid waste (MSW), collected in most developed countries; and samples of waste collection in urban areas, comparisons of gross national product (GNP), and solid waste generation per inhabitant for the rest of the world. Recent research suggests that the amount of MSW is strongly correlated to income level and lifestyle. Using the working definition of MSW and this estimation methodology, as of 2004 (the most recent data), the total MSW collected worldwide is projected to be 1.2 billion metric tons (Table 1.1).

How to dispose of the cans, cereal boxes, newspapers, tires, bottles, and other castoffs of communities in the industrialized and the developing world in an environmentally sound and economically efficient way has become a problem of critical proportions. Until recently, resources were considered as something scarce, which needed to be reused with little, if any, going to waste. To assist in this effort, 'rag men' and piggeries can be found in most urban areas of industrialized countries and form the basis of an active recycling industry [2].

Waste-to-Energy. DOI: 10.1016/B978-1-4377-7871-7.10001-2

Table 1.1 Estimate of Municipal Solid Waste Worldwide (2004)

Category	**Million metric tons**
OECD countries	620
CIS (Baltic States excluded)	65
Asia	300
Central America	30
South America	86
North Africa and Middle East	50
Sub-Saharan Africa	53
TOTAL	**1,204**

Source: Reference [1]

With population growth and waste generation rates spiraling upward, many communities worldwide have now begun to search for alternative long-term solutions to the methods they once employed to dispose of their solid wastes. Sanitary landfilling of solid waste has become the traditional approach in the industrialized world where it has progressed from an earlier era of dumps and open burning to its present state of engineered landfills [3].

These days, sanitary landfills can be designed to be an environmentally acceptable means of waste disposal, provided they are properly operated. However, new regulations regarding landfill liners, leachate control systems, landfill gas collection and control systems, and long-term closure requirements have dramatically increased the cost of 1andfilling. In addition, suitable land for landfill sites close to nearby urbanizing areas is now less available for many communities, thereby resulting in these communities having to locate more distant landfill sites. The Not-In-My-Back-Yard (NIMBY) attitude on the part of citizen opposition groups, however, has increased the difficulty of many communities in the siting and permitting of these new landfills.

Consequently, as existing landfill capacity has been reduced, there has been increased interest in the concept of recovering energy and recyclable materials from MSW rather than relying on sanitary landfilling as the primary long-term method of solid waste disposal. Further, the European Union (EU) goal to reduce landfilling by 65 per cent of biodegradable MSW and the EU Directives on Waste Incineration and Landfilling has prompted new construction of waste-to-energy (WTE) plants and upgrading of existing plants to meet EU Directives [4].

1.2 The trends towards WTE

Producing and utilizing energy from the combustion of solid waste is a concept that has been practiced in Europe since the turn of the last century. Prompted by concerns for groundwater quality, and the scarcity of land for landfilling, Japan and many European countries embarked on massive construction projects for WTE programs in the 1960s. Transfer of this technology to the United States first began in the late 1960s and early 1970s. In addition, other projects utilizing American technology in the area of shredded and prepared fuels were constructed. However, most of these projects were problematic, because they were unable to overcome materials handling and boiler operations problems. It was these failures that made local government leaders initially cautious in funding construction of WTE projects.

Nevertheless, several WTE projects were developed in the mid to late 1970s in communities such as Saugus, Massachusetts; Pinellas County, Florida; and Ames, Iowa; all of which were experiencing severe landfill problems. Success of these projects helped the WTE industry gain acceptance by local government leaders, and the financial community. Tax incentives made available by the federal government for WTE projects attracted private capital investment in such projects assisting in the maturing of this industry in the United States and sparked the development of many new projects [1].

At the time of writing, there are about 1,300 WTE facilities worldwide (Table 1.2), which are estimated to provide almost 600,000 metric tons per day of disposal capacity. Large numbers (Figure 1.1) are located in Europe (440), primarily because of the EU's directive that requires a 65 per cent reduction in the landfilling of biodegradable MSW. Nonetheless, a large part of the EU's waste stream (40%) is still landfilled. In 2009, these WTE plants converted about 69 million metric tons of MSW (or about 20% of the EU waste stream), generating 30 TWh of electricity and 55 TWh of heat [6]. This is roughly equivalent to supplying the annual needs of 13 million inhabitants with electricity and 12 million inhabitants with heat in these countries. Given the EU's directive on landfilling, estimates of new WTE facility construction range from 60 to 80 new plants by 2020. Scandinavian counties (Denmark and Sweden) have historically been significant proponents of WTE.

Asian countries (Japan, Taiwan, Singapore, and China) have the largest number (764) of WTE facilities worldwide. All of these countries face limited open space issues for the siting of landfills and have high urban populations. For example, Japan has addressed its solid waste issue by processing an estimated 70 per cent of MSW in WTE facilities.

One of the largest markets for WTE plant construction is in China. The Chinese WTE capacity increased steadily from 2.2 million tons in 2001 to nearly 14 million tons by 2007, although landfilling remains the dominant means of waste disposal in China. Since the beginning of the 21st century, this has made China the fourth largest user of WTE, after the EU, Japan, and the USA, with most plants located in the heavily industrialized cities in southeastern China. This is projected to increase to one hundred by 2012 according to the latest five-year plan. Despite the relatively high capital cost of WTE, the central government of China has been very proactive with regard to increasing WTE capacity. One of the measures brought in has provided a credit of about US$30 per MWh of electricity generated by means of WTE rather than by using fossil fuel [7].

In the United States, there are currently 87 WTE plants operating in 25 states managing about 7 per cent of the nation's MSW, or about 90,000 tons per day. This is the equivalent of a baseload electrical generation capacity of approximately 2,700

Table 1.2 Estimated worldwide WTE facilities

Country	Population million	Urban (%)	GNP/Cap. US$	MSW rate kg/day	Generated TPD	Capacity (%)	WTE plants TPD	WTE plants #
Macau	0.4	100		1.5	648	133	864	1
Hong Kong	6.7	95	22,990	1.7	10,821	0	–	–
Singapore	4.1	100	26,730	1.7	6,800	121	8,200	4
Taiwan	22.0	80	13,000	1.3	22,000	111	24,324	24
Korea	46.9	81	9,700	1.6	60,626	15	9,190	26
Thailand	58.2	20	2,740	0.5	6,053	6	390	2
Malaysia	22.7	54	3,890	0.8	9,874	0	–	–
Phillipines	74.3	54	1,050	0.5	20,941	0	–	–
China	1,238.0	30	620	0.8	296,340	2	5,650	7
India	1,010.0	27	340	0.6	173,235	0	–	–
Japan	126.6	78	39,640	1.5	144,415	138	200,000	700
Europe	300.0			1.3	390,000	59	228,850	440
France (included in Europe)	60.0		26,270	1.3	78,000	50	38,680	128
USA	263.1		28,020	2.0	526,200	17	91,453	96
Canada	29.6			1.8	53,280	3	1,800	2
World Total							**570,721**	**1,302**

Source: Deltaway Energy, Inc.

Figure 1.1 WTE facilities in Europe, 2008. (Cited in Reference [6]).

megawatts to meet the power needs of more than two million homes, while servicing the waste disposal needs of more than 35 million people.

During the 1990s, the WTE industry in the United States experienced a number of setbacks, which resulted in no new WTE facilities being constructed between 1995 and 2006 [5]. Expiration of tax incentives, significant public opposition in facility siting, and the US Supreme Court decision in Carbon dealing with solid waste flow control, forced many communities in the United States to opt for long-haul transport of their solid waste to less costly regional landfills. A subsequent Supreme Court decision on flow control restored the ability of communities to enact flow control ordinances and enable them to direct their wastes to WTE facilities. As a result, some WTE facilities began to expand by adding new processing lines to their existing operations. These facilities are basing their requests for financing and permitting on their successful records of operation and environmental compliance.

1.3 Climate change and WTE

WTE is internationally recognized as a powerful tool to prevent the formation of greenhouse gas emissions and to mitigate climate change. The International Panel on Climate Change (IPCC), the Nobel Prize winning independent panel of scientific and technical experts, has recognized WTE as the key greenhouse gas emission mitigation technology [8]. The World Economic Forum, in its 2009 report, *Green Investing: Towards a Clean Energy Infrastructure*, identifies WTE as one of the eight technologies likely to make a meaningful contribution to a future, low-carbon energy system [9]. In the EU, WTE facilities are not required to have a permit or credits for emissions of CO_2 because of their greenhouse gas mitigation potential. Further, the German Ministry of the Environment has projected that the application of the EU Landfill Directive will result in the reduction of 74 million metric tons of equivalent CO_2 emissions by 2016. WTE is designated as 'renewable' by the 2005 Energy Policy Act, by the United States Department of Energy (DOE), and by 23 state governments.

Over the years, there have been a number of quantitative assessments made to compare the environmental benefits associated with the processing of MSW in WTE facilities after recyclable materials have been removed rather than disposing of MSW in landfills. A state-of-the-art WTE facility is roughly estimated by most models to save CO_2 in the range of 100 to 350 kg CO_2 equivalent per ton of waste processed [10]. The variability is often expressed in such models as the differences in:

- Waste composition (% biogenic);
- Amount of heat and electricity supplied (i.e., the more energy supplied as heat the higher the CO_2 savings);
- The country/energy substitution mix.

Recently, markets have developed around the world to compensate WTE operators for the reduction in these CO_2 emissions. Currently, this CO_2 credit is higher in developing countries due to poor landfill practices. Further, the more efficient the WTE facility, the more CO_2 credit it will generate.

References

[1] UN-HABITAT. Solid Waste Management in the World's Cities. United Nations Human Settlements Programme 2009.

[2] Lacoste Elisabeth, Chalmin Philippe. World Waste Survey: Turning Waste into a Resource. Paris: Veolia Environmental Services; 2006.

[3] Eawag. Global Waste Challenge: Situation in Developing Countries. Swiss Federal Institute of Aquatic Science and Technology 2008.

[4] Manders Jan. Developments on Waste to Energy Across Europe. Confederation of European Waste-to-Energy Plants. Columbia: Presentation to WTERT; Jan 2010. October 8, 2010.

[5] America's Own Energy Source. American Society of Mechanical Engineers. Integrated Waste Services Association, Municipal Waste Management Association, and Solid Waste Association of North America; 1995.

[6] CEWEP. Confederation of European Waste-to-Energy Plants. Heating and Lighting the Way to a Sustainable Future, http://www.cewep.eu/information/publicationsandstudies/statements/ceweppublications/m_471; 2010.
[7] Themelis Nickolas, Zhang Zhixiao (2010). WTE in China. Waste Management World (August 2010).
[8] German Ministry of Environment, Waste Incineration – A Potential Danger? The Contribution of Waste Management to sustainable Development in Germany.
[9] Solid Waste Association of North America. Letter to Representatives Henry Waxman and Joe Barton, April 28, 2009.
[10] O'Brien Jeremy. Waste-to-Energy and Greenhouse Gas Emissions. MSW Management 2010.

2 Project implementation concepts

Chapter Outline

2.1 Introduction

The successful implementation of a WTE project rests primarily upon the following essential building blocks or key elements [1]:

- A reason or need for the project because of a critical community solid waste disposal problem or crisis;
- An implementing government agency or private project developer with political commitment willing and able to undertake the project;
- An adequate supply of solid waste for the project or means to assure waste stream control or attract sufficient quantities from other communities;
- Markets for the recovered energy and recovered materials; and
- A project site that is environmentally, technically, socially, and politically acceptable.

Waste-to-Energy. DOI: 10.1016/B978-1-4377-7871-7.10002-4

Perhaps the most critical element that must be in place if a WTE project is to succeed is that a need for the project exists. That is, a situation exists such that community leaders perceive that the community is facing an immediate or long-term solid waste disposal problem, and that planning for an alternative to sanitary landfilling should be undertaken.

A second major element that must be present for the success of the project is political leadership. Unfortunately, the most well conceived plans for public benefit projects are often implemented without such leadership. WTE projects are capital intensive and require planning that frequently extends over a two to five year period. Since most politicians are elected every two or four years, WTE projects can find themselves orphaned by new political leaders who may have different solid waste management agendas. Consequently, if a community has any hope of implementing a facility, then it is necessary to have an implementing entity (e.g., a county, municipality, authority, electric utility, waste hauler or contractor) or driving force, which has long-term political support.

The community must also be able to supply or divert enough solid waste to its proposed facility since solid waste serves as the feedstock for plant operations and energy production. The community must be able to guarantee both the quantity and quality of its solid waste. The level of the wastes to be guaranteed will determine the ultimate size of the facility.

Securing energy and materials markets is another critical component in implementing a WTE project. Such markets provide revenues that offset plant tipping fees and make WTE facilities financially attractive to both communities and private developers. Without these markets, WTE projects would not prove economically feasible for most communities.

Another critical component in project implementation is the securing of a site to construct and operate a facility. Siting public benefit projects, like solid waste facilities, has proven to be a time-consuming and controversial task in recent years. Such projects have attracted significant public opposition because of citizen concerns associated with perceived project impacts such as air quality, public health, traffic congestion, litter, noise, aesthetics, and property values. The Not-In-My-Backyard (NIMBY) attitude has caused project developers to spend more time in searching for project sites which are technically, environmentally, and socially acceptable.

What is somewhat unique about these projects, aside from their complexity, is the tortuous paths they have often taken from project inception through to construction. This lengthy implementation period for many projects has resulted in part from their inability to assemble an experienced project team having strong, long-term political support to move the project along when faced by major project impediments [2]. Without this strong management organization and support, many projects have been delayed by their inability to resolve critical problems, such as strong public opposition to a proposed site, the addressing of environmental concerns, and unfavorable project economics. While Chapters 3 through 10 will concentrate on the critical contractual, technical, and financial decisions that each project must successfully resolve, this chapter will primarily focus attention on developing the management structure in order to cope with these project issues on a day-to-day basis. It will also address the means to effectively communicate these complex issues to the public and the people overseeing the project.

2.2 Developing the project team

An extremely important aspect of assuring success of a program with the complexities of a WTE project is establishment of a strong project team that can guide such a long-term project to completion [3]. While every project is unique to some extent, there are a number of fundamental actions which government must be prepared to take to place a project on the right course for the long-term. Such measures can enhance the project's potential for success.

2.2.1 *Internal project team*

At the outset of a project, a key action is the establishment of an internal team that will have support from the political decision-makers [1]. Since such projects will require significant up-front development costs for staff and consulting services over several years, it is critical that there is a long-term commitment by the community to support the project. Without this commitment, it is unlikely that the project will truly ever succeed.

Ideally, the internal project team, which will direct the activities of the government agency's own staff and outside consultants or advisors, should have appropriate agency heads from their key administrative, public works, financial, legal, environmental, and communications areas [4]. The purpose of this interagency committee is to guide the project through key decision points and to provide policy recommendations to the political decision-makers. Since the agency heads of all key government departments are members of this project committee, it is more likely that project decisions will receive a more balanced and thorough review before presentation to the elected representatives within the community, and result in unified staff recommendations [5]. One disadvantage of this type of project structure, however, is that it requires extensive time commitments on a continual basis from government departments, which may be unable, because of their own project demands and budgets, to supply these services. Thus, many projects have been unable to organize such an interagency management committee in practice.

To compensate for a lack of full-time support from outside agencies, many solid waste, public works, planning, or utility departments responsible for implementation of their community's projects have established the position of a full-time project manager. They have recognized that the coordination of their government's staff and outside consultants and advisors was essential to the success of the WTE project. The person selected for this position often comes from within a particular government department having responsibility for the project, or is hired from outside government as a contract employee. Regardless of this individual's civil service status, his or her role typically is to be responsible to coordinate, schedule, and monitor the activities of the internal project team and consultant staff. Other full-time staff members often assist this person, because the time required to successfully undertake these roles can be very significant.

2.2.2 Consultants and advisors

An independent consulting team with an excellent track record in WTE implementation should be hired at the outset of the project to complement the government's internal project team. Many governments hesitate to utilize consultants at early phases in such projects because of the significant costs associated with obtaining consulting services. Unfortunately, this is shortsighted because consultants are usually cost effective in the long-run as they will add credibility and needed expertise to a community's project.

An independent consulting engineering firm, which has significant experience in the WTE area, can provide the community with valuable insights. The feasibility report that it prepares can point out the advantages and disadvantages to the community with respect to such key issues as project ownership, financing, procurement, siting, and permitting. If nothing else, its assistance in drafting the project's request-for-proposal should enhance the quality of this procurement document, which can result in more responsive vendor proposals to the community's needs.

Since the bond issues for most WTE projects are usually the largest ever issued by most local governments, it is also important that a strong outside financial team is selected to prepare the overall financing for the project. This team usually includes one or more bond underwriting or investment banking firms, an independent financial advisor, a bond counsel, and a bond underwriter's counsel. Each of these firms or individuals has a specific role in helping develop a strong financing plan for the project which will be favorably viewed by bond rating agencies and the credit markets. Chapter 10 describes some of these different roles in detail.

As the project nears the procurement phase, some governments may add other experts to the consulting team. For example, an insurance advisor is sometimes hired to develop the technical insurance requirements for the Request-for-Proposal or Tender, and to assist in securing insurance coverage for the project. In addition, special legal counsels are often retained to: assist in environmental permitting; help negotiate energy sales contracts; assist in the preparation and negotiation of construction and operations contracts with the selected contractor for the facility.

2.3 Risk assessment

A key component of the feasibility analysis for a WTE project is an assessment of the possibility that a single or multiple events, which will have a detrimental impact upon a project, might occur, and who should bear this loss [6]. Many of the decisions that a community makes along the path of project implementation are concerned with allocation or assignment of risk events in the following general areas: waste stream control; energy markets; legal and regulatory arrangements; project construction; and project operation. While it is difficult to determine the probability of each particular risk or exposure, which usually results in the monetary loss, such events can be categorized by who was responsible for the cause of the event (i.e., the community, contractor, or force majeure). Table 2.1 lists the major risk events in the project and the typical assignment among project participants.

Risk sharing among project participants is usually achieved through the negotiation of contracts. While any party can assume a project risk, it is more typical in WTE projects for the party responsible for controlling the cause of a potential loss to be allocated that risk. Assumption of risk beyond the reasonable control of a party usually requires that they are adequately compensated for such risk sharing. Thus,

Table 2.1 Typical assignment of project risks

	Procurement approaches			
			Full service	
Risk event	**A/E**	**Turnkey**	**Public**	**Private**
Waste stream:				
Failures in waste stream	Government	Government	Government	Government
Changes in waste composition	Government	Government	Government	Government
Energy and materials markets:				
Energy and materials revenues	Government	Government	Government	Government
Legal and regulatory:				
Tax law changes	Government	Government	Government	Shared
Environmental permitting	Government	Government	Government	Shared
Anti-trust challenges	Government	Government	Government	Government
Facility construction:				
Design errors	Government	Contractor	Contractor	Contractor
Equipment performance	Contractor	Contractor	Contractor	Contractor
Costs underestimated	Shared	Contractor	Contractor	Contractor
Failure of subcontractor	Shared	Contractor	Contractor	Contractor
Subsurface conditions	Government	Government	Government	Government
Inflation	Either	Either	Either	Either
Strikes	Either	Either	Either	Either
Force majeure	Government	Government	Government	Government
Facility operation:				
Plant performance	Government	Government	Contractor	Contractor
Damage by waste	Government	Government	Government	Government
High inflation	Government	Government	Government	Government
Operating costs	Government	Government	Contractor	Contractor
Residue disposal	Government	Government	Government	Government

Source: Reference [6].

while full-service contractors can be expected to assume greater project risk than contractors either in an architect/engineer (A/E) or turnkey procurement approach, there is usually a direct correlation between the level of such risk and the overall construction price of the WTE project. This is due to the nature of the project performance and corporate financial guarantees that a full-service contractor must assume. Consequently, it is critical that a risk assessment be made early in a project so that certain potential project events are understood and a risk and compensation posture is developed.

2.3.1 Waste stream

Historically, any risks associated with assuring that a reliable supply of solid waste is delivered to a facility, are assumed by the community. This means that a community must be able to fully guarantee that solid waste generated within its boundaries will be delivered to its proposed WTE facility. Waste stream control can be achieved through a number of methods, such as the use of state legislation or local ordinance, long-term put-or-pay contracts, subsidized tipping fees, or governmental collection of solid waste.

In addition to guaranteeing delivery of solid waste to a WTE facility, the community must usually assume the risk of the quality of such waste – that is, the heating value or Btu content of the waste, its percentage of moisture, and percentage of combustibles. The task of identifying the plant's reference waste composition is discussed in more detail in Chapter 4. The community typically guarantees a reference solid waste composition that a contractor assumes for design purposes. Thus, if this reference composition changes, perhaps due to increased recycling of paper products, then the community must assume the responsibilities and costs of this change, generally through an increase in tipping fees.

2.3.2 Energy and materials market

The risks associated with energy and materials markets are typically assumed by a community. In the event of lower prices for energy and recycled products than anticipated, the community is usually responsible for subsidizing project revenues through increased tipping fees. These risks may be partially mitigated by securing negotiated long-term contracts with the energy or materials customers.

2.3.3 Legal and regulatory

Risks associated with changes in laws and governmental regulations, which are generally unforeseen or uncontrollable, are generally allocated to a community. It is difficult at the outset of a project to anticipate changes in critical federal or state legislation and rules in areas such as tax law, environmental protection issues, and antitrust challenges in the courts. Through negotiation with a full-service contractor, which desires to own the community's facility, these tax-law risks can often be allocated to the contractor.

2.3.4 Facility construction

Risks associated with construction of a WTE facility include the following: design errors; strikes; failure of subcontractors to perform; equipment performance; cost underestimated; subsurface conditions; inflation; and force majeure. Typically, those risks associated with the performance of the design, technology, or equipment are the contractor's responsibility. Furthermore, risks within the contractor's reasonable control are typically his or her financial responsibility. Contractors for WTE facilities will generally guarantee construction price (subject to normal inflation), the length of construction, and the performance of any subcontractors to be on time and within budget; these guarantees are generally supported through construction and performance bonds, project insurance policies, and incentives by a community to encourage timely completion. The exception is in the case of the A/E procurement approach where these risks are typically shared with the community through negotiation. For example, a maximum contract price could be assigned to a contractor under the 'chute-to-stack' procurement approach.

Another cause of risks associated with project construction is force majeure, or uncontrollable events. These risks can include Acts of God, such as floods, earthquakes, or other natural disasters; sabotage; war; explosions; and unforeseen site conditions, such as subsurface soil conditions. Historically, these risks are assumed by government, which can mitigate these losses by procuring special insurance for the project.

2.3.5 Facility operation

Risks associated with the operation of a WTE facility include: continued plant performance; damage to the plant by the community's solid waste; inflation; operating costs; and residue disposal. Under both the A/E and turnkey procurement approach, a community assumes all the risks of continued performance and operation costs of the project after its warranty period. In contrast, the community has the option to share some of these risks or pass them on to a contractor under the full-service procurement approach. Historically, full-service operators will guarantee specified plant performance levels, and maximum annual operating and maintenance expenses. However, these guarantees are sometimes subject to occurrences beyond the control of the contractor, such as: above normal inflation; damage to the plant by hazardous or explosive materials contained in the solid waste stream; or inability to dispose of ash residues to a legally permitted landfill. Such risks are generally assumed by the community.

2.4 Implementation process

Implementation of a WTE project is a complex process that consists of several phases requiring 'go/no-go' decisions to be made by the project participants [7]. Since no two WTE projects are identical, the discussion on project implementation in this section is

Table 2.2 Typical implementation phases of a WTE project

Phase I – Feasibility analysis	
• Waste stream analysis	Review permitting requirements
• Waste disposal practices analysis	Risk and legal assessment
• Energy and materials market study	Financial analysis
• Analysis of feasible WTE technologies	Develop project alternatives
• Analysis of potential facility sites	Go/no-go decision
Phase II – Intermediate phase	
• Securing waste stream	
• Finalize energy market contracts	
• Develop environmental permits	
Phase III – Procurement	
• Select project alternative	Develop financing plan
• Select site and acquire	RFQ/RFP produced and issued
• Permitting underway	Contractor selected
• Market contracts concluded	Contract negotiations concluded
• Waste stream guarantee	Notice-to-proceed
Phase IV – Plant construction	
• Site preparation	Equipment installed
• Complete final design	Testing and startup
• Equipment ordered	Acceptance testing
• Building constructed	Certificate of Completion
Phase V – Plant operations	
• Service fee payment	Annual report (optional)
• Annual tipping fee adjustment	Facility retesting (optional)

by nature generic in approach. Table 2.2 (above) displays the steps normally followed in the implementation process for a typical WTE project. These steps can be modified by a community taking into account its individual concerns and needs. The critical point is that a well-planned implementation process is essential to an expeditious, cost-effective, and successful project.

2.4.1 Project phases

2.4.1.1 Phase I – Feasibility analysis

This phase is preceded by a preliminary, and somewhat informal, investigation by a community or project sponsor that the major project elements (as discussed above) exist, and make further study of a potential WTE project reasonable.

During Phase I, the feasibility of such a project is evaluated in detail. An analysis is undertaken of the community's existing, and projected, waste stream to determine the ultimate size of a single or multiple unit WTE system. Markets are examined to determine whether the energy and/or materials produced by this system can generate adequate revenues to offset the construction, operating, and financing costs for the facility. Feasible sites are investigated along with an analysis of the technical, environmental, and institutional requirements for permitting a facility. In addition, a risk assessment is undertaken; this helps to determine the risk posture to be taken by the community regarding WTE technology, project ownership, operation, and financing. A formal feasibility report, which is then presented to the implementing entity, usually documents all activities in this phase. Phase I ends with a 'go/no-go' decision on the part of the community to either proceed with implementation of the project or to terminate activities for the foreseeable future.

2.4.1.2 Phase II – Intermediate phase

This phase encompasses all of the intermediate steps leading up to the procurement of the WTE system. This includes securing the put or pay contract for the waste stream, the energy market contract, and preliminary environmental permits. Typically, many WTE owners and operators initiate their public education program and develop a concept on plant architecture and visitor centers.

2.4.1.3 Phase III – Procurement

Phase III incorporates all the steps necessary to procure the WTE system desired by the community, including: contracts for waste supply, energy and materials markets, plant construction and operation (if applicable); acquiring the project site; obtaining all environmental permits and/or regulatory approvals; and securing the financing for the project. Phase III usually requires the development of specialized procurement documents such as a Request-for-Qualifications (RFQ) and Request-for-Proposal (RFP). Dependent on the financing approach selected by the community, documents such as a Bond or Trust Indenture, Engineer's Feasibility Report, and Bond Prospectus may need to be written.

2.4.1.4 Phase IV – Plant construction

Phase IV covers all the steps necessary in the construction of the WTE facility, including: site preparation; final design; ordering facility equipment; installing this equipment; testing and startup of the facility; and completing acceptance tests. The selected facility contractor usually undertakes these tasks, although most communities obtain the services of a knowledgeable engineering consulting firm to independently monitor construction activities and the facility's acceptance tests. Phase IV usually ends with the facility successfully completing the contractual acceptance test procedures, and the community issuing a Certificate of Completion to the contractor.

2.4.1.5 Phase V – Plant operations

Phase V covers the period from plant acceptance through the length of the operations period, which may last 20 years or more in the case of a full-service contractor for a publicly- or privately-owned facility. These contracts usually have clauses allowing them to be renegotiated after this time period. In contrast, operations contracts for turnkey operators can be significantly shorter, generally five years or less.

During this period of plant operations, communities may have a varying degree of project responsibility. In the case of governmental operations, they have the full range of operational responsibilities. However, in the case of a privately owned and operated facility, the community may only have the right to request that the facility is retested to prove that its original performance guarantees can be met. Most communities retain independent consulting engineering firms to undertake these oversight roles.

2.5 Implementation project scheduling

The duration of the implementation process for a WTE project depends upon a variety of factors that project sponsors may or may not control – factors such as procurement approach, delays in regulatory reviews, and length of negotiations for contracts; for example, the length of time that a community may need to conclude the implementation of a project will have an impact. Consequently, it is difficult to provide a rule of thumb for project duration that will apply to all projects. Generally, however, a full-service procurement approach in WTE projects can be completed faster than either the A/E or turnkey approaches. This is primarily because the community can simultaneously undertake all the remaining procurement activities on a fast-track basis with limited project delays.

With significant numbers of WTE projects implemented in recent years, it is expected that the length of time required to execute such projects can be reduced. Early projects required upwards of two to three years to complete planning and procurement activities, with an extra two years to complete all contractual negotiations and arrange for financing. This was in addition to the two to three years required for design and construction of the facility. Thus, many communities desiring to implement a WTE project had to look accept an extended waiting period before their facility was fully completed.

Current experience suggests that WTE projects can be implemented in a somewhat shorter space of time. It appears that, as feasibility procedures become more refined, Phase I activities can be completed in an average of six months, followed by Phase II procurement activities lasting an additional 12 to 18 months. Once contracts are concluded, and financing arranged, design and construction activities can be completed in an average of 24 to 36 months. Some projects have even further reduced these schedules by fast-tracking procurement, financing, and construction activities.

2.6 Implementation project costs

Because they are highly capital intensive, WTE projects require significant amounts of upfront expenditures on the part of sponsors for implementation activities. Activities on the part of staff and consultants often require expenditures depending on the complexity of the WTE project. The activities include: consulting engineering services; legal, financial and insurance advice; land acquisition; subsurface soil testing; land surveys; payment and performance bonds; bond insurance; liability and force majeure insurance; bond ratings and credit reports; printing of official statements and bonds; miscellaneous bond issuance expenses; and expert witnesses.

2.7 Public information programs

An effective public information program is essential to the success of any WTE project. The public information program can be, in many respects, similar to that used in most public benefit projects. For example, details concerning the project can be disseminated to the general public through the following typical media: periodic project newsletters; brochures; press releases; movies; slide shows; visual displays; press packets; radio and television interviews by project and staff consultants; presentations before influential civic, fraternal, business, and environmental organizations; and formal public workshops and hearings.

Typical public benefit projects differ from WTE projects as they generally center on the nature and complexity of issues that concern citizen groups. Historically, such projects have attracted significant public opposition due to typical concerns such as: public health and safety; traffic congestion; noise; litter; aesthetics; and property values. Project sponsors must address these public concerns early during implementation in order to minimize the dissemination of information about the project to the community [8]. However, this slows down the project's public approval process. Consequently, it is essential that information and involvement programs is developed early in the process, to help elicit public input regarding project decisions, and to respond to public concerns in a timely, but thorough manner. These approaches have historically paid dividends for many projects, particularly during the period when a project siting decision has its greatest opposition, allowing successful implementation. A number of different strategies have been utilized successfully.

Some project sponsors have presented detailed information regarding the project in informal and formal public forums. Project staff record specific citizen concerns expressed at these meetings. Where additional technical expertise is needed in the project team to effectively respond to these issues, nationally recognized experts are retained. Subsequent to these meetings, a well-researched response to each of the technical issues raised is presented back to these groups. This may result in modifications or compromises in project recommendations. In this way, the project team is able to successfully defend its recommendations, keep the political process moving along, and win public support.

Other project sponsors have utilized the services of an independent citizen review committee or task force, composed of representatives of various community groups. The purpose of this committee is to provide a forum for resolving issues of public concern regarding the project. These groups have worked well when they receive explicit charters and specific timeframes from the project sponsor.

Such committees provide a means of dialog between the project sponsor and a group representing community interests to discuss technical, economic, and social issues. This forum offers project sponsors the opportunity of responding to citizen concerns by making accommodations – a process that would otherwise be missing in the formal permitting process. By beginning this dialog prior to the start of permitting activities, project sponsors can often have a better chance of avoiding significant implementation delays.

References

[1] Nemeth Diane M. The Resource Recovery Option in Solid Waste Management: A Review Guide for Public Officials. Chicago: American Public Works Association; 1981. DOE/CS/20156-T1.

[2] Schoenhofer Robert F. Project Structure-Process and Players. In: Proceedings of The Third Annual Resource Recovery Conference. Washington: National Resource Recovery Association; 1984. Washington DC, March 28–30.

[3] Rose David P. Project Structure-Process and Players. In: Proceedings of the Third Annual Resource Recovery Conference. Washington: National Resource Recovery Association; 1984. Washington DC, March 28–30.

[4] Berry Patricia V, Rogoif Marc J. Teamwork Plus Communications Equal Success. Waste Age 1986. November: 105–108.

[5] Scaramelli Alfred B. Resource Recovery Success Depends Upon Commitment. American City and County 1984. May: 30–34.

[6] Zier Robert E. Managing Risks Part of Success. Solid Wastes Management 1982. May: 35–39.

[7] US Environmental Protection Agency. Resource Recovery Management Model – Overview. Washington: US Environmental Protection Agency; 1980. SW-768.

[8] The Keystone Siting Process Group. The Keystone Siting Process Handbook. Austin: Texas Department of Health; 1984.

3 WTE technology

Chapter Outline

3.1 Introduction

One of the first questions an agency must answer is what technology will be chosen to convert its solid waste into energy. Each agency must identify and evaluate the various WTE technologies that are available and make its own selection based upon the requirements specific to its particular project. This includes consideration of factors (which will be discussed later) such as: available energy and materials markets; the size of the community's waste flow; site availability and location; capital and operating costs; ownership and financing considerations; and the level of risk to be assumed by the community or the facility operator.

Waste-to-Energy. DOI: 10.1016/B978-1-4377-7871-7.10003-6

In evaluating whether or not one technology better suits its needs than another, a community may often discover that one or more of their goals established for the project may conflict with others. A particular technology, for example, may produce the greatest amount of energy for the community's waste, albeit at the highest projected capital and operating costs. The selection of a technology, therefore, is not a simple one, but one that can require tradeoffs between one agency's goal with others. Since the risks associated with WTE technology can be substantial, it is critical that each community attempt to minimize these risks as best it can. The following criteria can be utilized to assess the relative risk of a particular WTE technology:

- Degree and scale of operating experience – Technologies have only been proven in pilot or laboratory operations, or with raw materials other than municipal solid waste. Other technologies have only been commercially operated in small facilities and the scale-up to larger sized plants may result in unforeseen problems;
- Reliability to dispose of municipal solid waste – The technology selected must be capable to dispose of solid waste in a reliable manner without frequent mechanical downtimes resulting in diversion of such waste to landfills;
- Energy and material market compatibility – The technology must be capable of recovering energy and materials for which markets are available;
- Environmental acceptance – The technology must meet all permitted environmental requirements established by regulatory agencies; and
- Cost to the agency – The technology must dispose of the community's solid waste at a price the agency can afford given alternative means of disposal.

3.2 Basic combustion system

The combustion of solid waste is accomplished in a furnace equipped with grates. A solid waste combustion system with energy recovery includes:

- Some type of structure to house the furnace and its appurtenances;
- A 'tipping floor' where the solid waste from collection and transfer vehicles is deposited;
- A storage pit or floor to store the solid waste delivered (solid waste combustion is a seven day per week, 24 hour per day operation; storage space is provided to enable this continuous operation);
- A charging system (normally overhead cranes) which mixes the various solid wastes received to develop a somewhat uniform material and then lifts it from the storage pit or floor and feeds (charges) the furnace;
- One or more furnace subsystems (sometimes referred to as combustion trains) which receive and burn the solid waste;
- A grate unit to move the solid waste through the furnaces; the most common grate designs are:
  - Reciprocating grates – This grate design resembles stairs with moving grate sections which push the solid waste through the furnace
  - Rocking grates – This grate design has pivoted or rocking grate sections which produce an upward and/or forward motion to move the solid waste through the furnace
  - Roller grates – This grate design has a series of rotating steep drums or rollers that agitate and move the solid waste through the furnace;

- Air pollution control subsystems to clean up the combustion gases; and
- An ash handling subsystem to manage the fly ash and bottom ash produced from the combustion of solid waste.

In Table 3.1 are listed some typical plant systems oftentimes specified for mass burn, WTE facilities.

Table 3.1 List of typical mass burn WTE equipment

Refuse		
1	RW	Refuse receiving
2	RB	Refuse bunker
3	CR	Crusher
4	RC	Refuse handling (cranes, hopper, chutes)
5A	WO	Recycled oil system
5B	LT	Leachate handling and treatment
Combustion		
6	IN	Incineration units
7	RE	Refractory, insulation, ports and doors
8	HY	Hydraulic system
9	IA	Primary and secondary air, air heater, ducting
10	FO	Fuel oil system
Boiler		
11	BO	Boiler
12	BS	Boiler structure, walkway and access
13	SB	Soot blowers
Air pollution control		
14	DN	De-NO_x system
15	LM	Lime system
16	AC	Activated carbon system
17	FG	Flue gas treatment system (cooling, ducting, etc.)
18	BH	Baghouse
19	ID	ID fan
Ash system		
20	BA	Bottom ash treatment system (extractor, conveying, etc.)
21	BC	Bottom ash crane
22	FC	Fly ash conveying system, silos
23	FA	Fly ash treatment system
Water and steam		
24	CW	City water
25	DW	Make up water treatment system
26	CD	Chemical dosing and conditioning
27	FW	Feed water system
28	SC	Steam distribution and condensate system
29	EC	Equipment cooling water system
30	AP	Auxiliary pumps
31	PA	Piping, valves and associated equipment
32	IS	Insulation

(Continued)

Table 3.1 List of typical mass burn WTE equipment—Cont'd

Power generation		
33	TG	Turbo generator
34	CO	Condenser
Electrical		
35	HV	High voltage electrical system, transformer
36	LV	Low voltage electrical distribution system
37	DP	Emergency power supply
Control and monitoring		
38	CS	Control and monitoring system
39	CE	Continuous emission monitoring system (CEMS)
40	CV	Closed circuit television (CCTV)
Civil part		
41	CI	Civil and architectural works
42	CH	Chimney
43	LF	Lift
44	RM	Rooms
45	AV	Air conditioning and ventilation (HVAC)
46	LI	Lighting and electric receptacles
47	FF	Fire alarm and fire fighting system
48	TC	Telephone and communication equipment
49	PM	Plant model and visitor facilities
50	MI	Miscellaneous works at plant area and outbuilding
Other systems and facilities		
51	WW	Wastewater treatment system
52	CA	Compressed air system (service and instrument air)
53	OP	Odor prevention
54	TW	Truck washing facilities
55	HC	Maintenance hoists and cranes
56	SP	Spare parts
57	DO	Documentation, drawings, and manuals
58	WS	Workshops
59	LA	Laboratory

Source: Deltaway Energy, Inc.

3.3 Stages of combustion

Solid waste normally has a moisture content of 20–25 per cent by weight. In order to successfully burn solid waste in a furnace, this moisture must be evaporated. Generally, most solid waste combustion units have three stages of reaction:

- Drying – Moisture driven off;
- Ignition – Solid waste ignited;
- Burnout – Solid waste is gradually moved through the furnace by the grate subsystem where the combustible organic fraction of the solid waste is burned out.

Successful combustion of solid waste is accomplished by controlling the '3 Ts of Combustion' – Time, Temperature and Turbulence.

- Time – The period taken for solid waste to pass from the charging hopper until the bottom ash is discharged at the end of the grate subsystem (usually 45–60 minutes);
- Temperature – Usually exceeds 1,800°F (982°C) within the furnace and is directly proportional to the residence time. If there is insufficient time in the furnace, the combustion reaction cannot proceed to completion and temperature declines; and
- Turbulence – Provided by the grate subsystem moving the solid waste downward through the furnace to expose it to and mix it with air.

Normally, solid waste combustors reduce the original weight of the solid waste by 75+ per cent and the volume by 85 to 90 per cent.

Combustion is aided by the introduction of air at two locations in the furnace. Air is introduced underneath the grates (underfire air) to increase the agitation and turbulence within the furnace and help cool the grates. Air is also introduced above the burning solid waste (overfire air). Overfire air ensures that there is adequate oxygen available to completely oxidize and burn the entire combustible fraction of the solid waste. Overfire air also aids mixing of the combustion gases, thereby ensuring complete oxidation and destruction. Combustion gases (also called flue gases) move from the furnace through the flues and the air pollution control systems, and are eventually discharged out of the stack into the atmosphere.

3.3.1 WTE solid waste combustors

In a WTE solid waste combustor, the energy released from combustion in the form of heat is used to generate steam in a boiler. The common method of capturing this released energy is either through refractory or waterwall furnace systems. The major difference between these two designs is the location of the boiler.

- Refractory units – This design consists of boilers located downstream of the combustion (furnace) chamber. The hot combustion gases pass through the boiler tubes to create steam;
- Waterwall units – This design has the furnace constructed with water tube membrane walls to recover the heat energy directly from the furnace unit. Waterwall designs are more commonly used because their thermal efficiency is higher than that with refractory units.

Boilers convert the heat released to steam, which can be used either to generate electricity or for industrial steam, applications (if a customer is nearby). Turbine-driven generators driven by the steam, generate electricity.

Combustion chamber design varies in geometry to accommodate to waste heating value and air starving conditions. Figure 3.1 illustrates these geometry options.

Table 3.2 describes the pros and cons of the most common geometries.

3.3.2 Products of combustion

Other than the release of energy in the form of heat, the products of combustion of solid waste are fly ash and bottom ash.

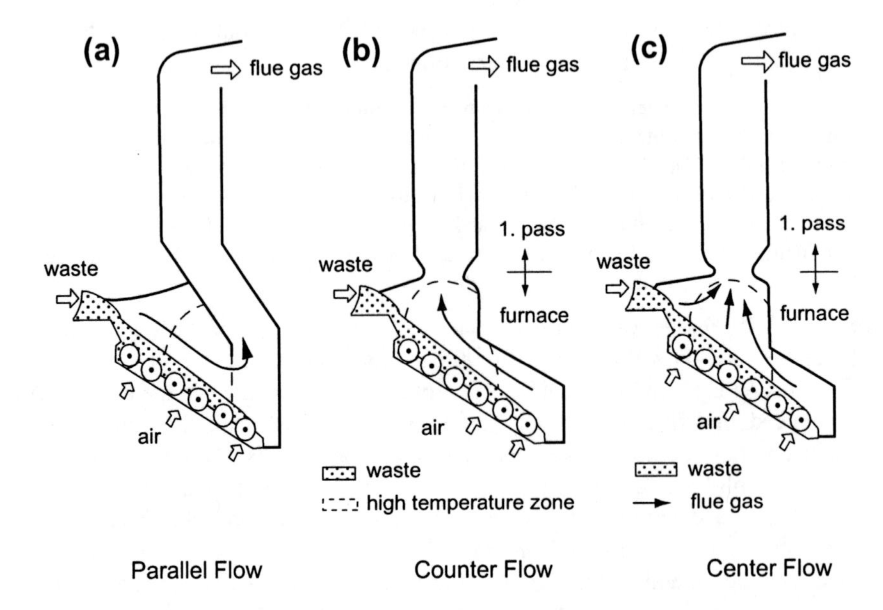

Figure 3.1 Furnace geometries (schematic) for municipal waste incineration plants.
Source: Deltaway Energy, Inc.

Table 3.2 Advantages and disadvantages of different furnace geometry designs

	Parallel flow	Counter flow	Center flow	Primary and secondary combustion chambers
Advantages	Pyrolysis gases pass through hottest area providing high gas burn out.	Energy transfer from the main combustion area to the drying and gasification area.	Very flexible for variable heat release distribution across the grate area.	Modular construction, reduced air flow in primary chamber, reduced fly ash.
	Suitable for waste with low calorific values.	Suitable for waste with high calorific values.	Suitable for waste with a wide range of calorific values.	Suitable for waste with high calorific values.
Disadvantages	Energy transfer from main combustion area to ignition area is provided by radiation.	Pyrolisis gases could bypass the hottest areas and may cause gas burn out problem.	Flow and mixing pattern into the first path chamber are sensible to disturbances.	Risk of poor combustion and low bottom ash burnout.

Fly ash is carried in the combustion gas, which also contains a number of contaminants, including acid gases, and other products of incomplete combustion. The gases pass through a variety of air pollution control devices for cleanup before being discharged out of the stack into the atmosphere.

Bottom ash is the non-combusted material, which is discharged at the end of the grate subsystem. As it is discharged from the grates, the bottom ash is still burning and is normally quenched by water. In the United States, the two ash streams (fly ash and bottom ash) are normally combined for management. The two combined ash streams are commonly referred to as solid waste combustor ash, or simply ash. In Europe, these two ash streams are not usually combined, and are normally managed separately.

3.4 Mass-burning

'Mass-burning' refers to the generic name for the type of technology used to incinerate unprocessed solid waste, and thereby release its heat energy. The thermal reduction of solid waste through mass-burning is now a commonly accepted procedure used throughout the world. There are decades of experience in constructing and operating some 500 mass-burn facilities in the United States and Europe. In Hamburg, Germany, such facilities were in operation as early as 1896, converting solid waste into electricity.

During the period from about 1905 through 1945, there were many overall improvements to these mass-incineration systems. The traditional incinerator constructed at this time had boilers that were refractory-lined to protect the outer shells of the boiler from sudden changes in temperature. Excess combustion air in the range of 100 to 200 per cent above combustion requirements was also generated in such units to further cool the walls of the boiler. Unfortunately, the large quantities of excess air produced by these facilities affected the levels of pollutants emitted by the refractory-lined incinerators. By the time the Clean Air Act was passed in 1970, this type of incinerator had fallen into disfavor, and many existing facilities were abandoned rather than retrofitted to meet the more stringent air pollution control requirements.

Refractory-lined incinerators also had the additional problem of being less efficient in recovering significant quantities of energy from solid waste. The energy was recovered in the combustion process of such incinerators centered in a waste-heat boiler located downstream from the combustion chamber. The large volume of exhaust gases from combustion pass through this boiler where the heat is absorbed and turned into steam. In such cases, heat recovery is much less efficient than if the boiler were located closer to the combustion chamber.

In the post World War II era, some of the European stoker manufacturers began to experiment with replacing the refractory material in the boilers with waterwall tubing for greater heat transfer. This allows steam to be produced at greater temperatures and pressures. In such incinerators, the walls of the furnace are lined with tubes filled with water, thus the name waterwall incinerators. The world's first integrated waterwall incinerator began operating in Bern, Switzerland, in 1954. It is still in operation today.

The impetus for this interest in energy recovery from solid waste was the rising cost of energy throughout the industrialized world.

3.4.1 Process description

An illustration of a typical mass-fired, WTE facility is shown in Figure 3.2. Solid waste collection and transfer vehicles proceed into a tipping area where they discharge their waste into a large storage pit, large enough to allow two to three days' storage or stockpiling of refuse. This allows plant operations to continue over weekends and holidays when deliveries are not accepted. There are some facilities that differ in design by utilizing a tipping floor with a front-end loader and belt conveyor system as their form of storage and feed system. However, in almost all facilities the refuse is fed into the furnaces by means of overhead cranes manipulated by a crane operator. Much of the success of the operation depends upon the skill of the crane operator to remove large or unusual objects in the waste stream that would otherwise prove to be a problem if fed into the boiler. The operator is also responsible to observe the nature of the incoming waste so that materials with different moisture contents are gradually intermixed and achieve uniform moisture content.

The refuse is then discharged into refuse feed hoppers, which meter out the refuse into the combustion chamber, either by gravity feeding or by a hydraulic feeding device. In the majority of systems, the waste is then pushed onto an inclined, step-like, mechanical grate system that continuously rocks, tumbles, and agitates the refuse bed by forcing burning refuse underneath newly fed refuse. Generally, most systems have three zones of activity along the grates: drying, ignition, and burnout. Holes in each grate bar allow underfire air to pass through the grates, resulting in cooling, and thus preventing thermal damage to the grate system. The width of the grate and the number of grate steps is dependent not only upon the manufacturer's specifications, but also on the overall size of the WTE system. There are five basic moving grate designs:

- The reciprocating grate – This grate resembles stairs with alternating fixed or moving grate sections. The pushing action may be in the direction of waste flow or in an upward motion against the waste flow;
- The rocking grate system – Pivoted or rocked grate sections produce an upward or forward motion, advancing the waste down the grate;
- The roller grate – A series of rotating stepped drums or rollers agitate the waste and move it down the grate;
- The circular grate – A rotating annular hearth or cone agitates the waste; and
- The rotary kiln – As an inclined cylinder rotates, it causes a tumbling action to expose unburned material and advance the waste down the length of the kiln.

Facilities using mass-burn technologies have been designed with either refractory or waterwall furnace systems. The major difference between these systems is the location of the boiler. Refractory units have their boiler located downstream of the combustion chamber, whereas waterwall is constructed with water tube membrane walls to recover the heat energy. A majority of mass-burn facilities constructed have

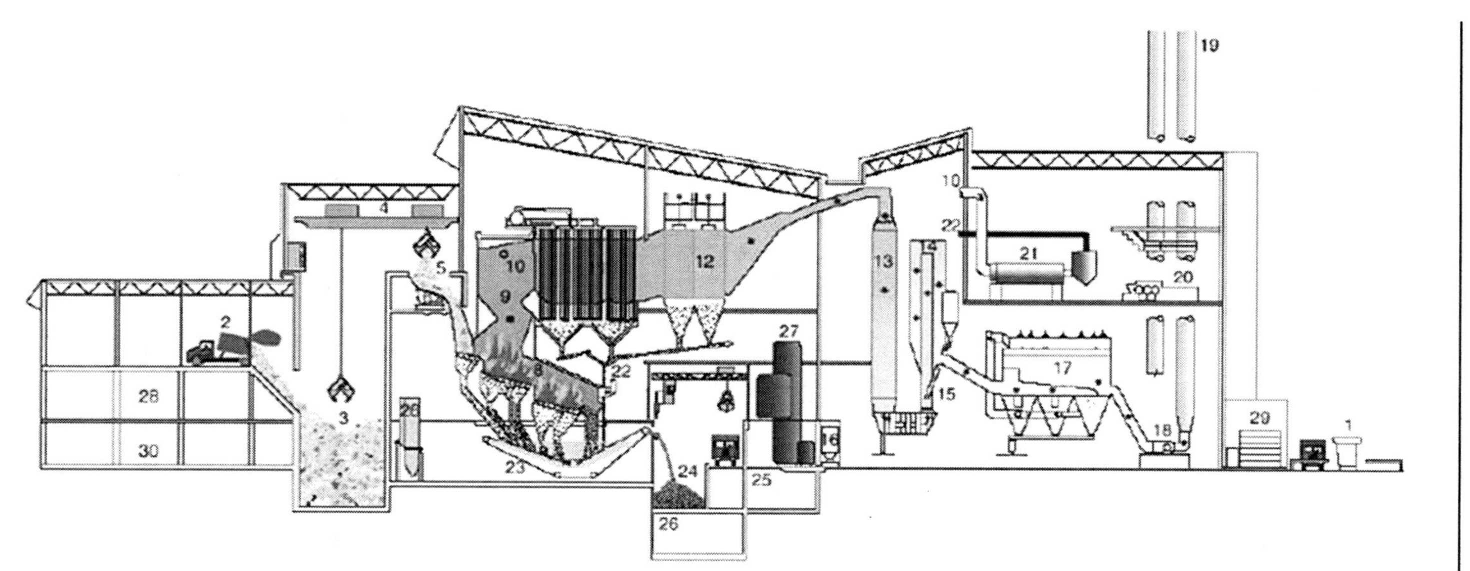

Figure 3.2 Example of cross-section of a mass-fired waterwall facility. Legend in English: 1, Scales; 2, Tipping floor; 3, Waste storage pit; 4, Cranes; 5, Feed chute; 6, Feed table; 7, Incineration grates; 8, Combustion chamber; 9, Secondary air injection; 10, First path; 11, Boiler; 12, Electro precipitator; 13, Flue gas cooling tower; 14, Scrubber; 15, Lime injection; 16, Activated carbon metering bin; 17, Baghouse; 18, ID fan; 19, Stack; 20, Sludge filter press; 21, Sludge dryer; 22, Dryer gas exhaust; 23, Ash extractor; 24, Bottom ash pit; 25, Wastewater tank; 26, Wastewater treatment; 27, Boiler feedwater purification; 28, Maintenance shops; 29, Dry sludge loading; 30, Spare parts storage.
Source: Deltaway Energy, Inc.

waterwall systems because of their greater thermal efficiency, which is generally between 60 and 75 per cent.

In modern waterwall incinerators, proper combustion of the waste is achieved through the introduction of air at two locations in the furnace. One location introduces air underneath the grate system (underfire air) to cause agitation and turbulence within the burning waste, and to help cool the grates. Air is also introduced above the burning waste (overfire air) to ensure that there is adequate oxygen available to completely oxidize and burn all the combustible materials, as well as promote proper mixing of combustion gases. During the combustion process, flue gases, heated to temperatures as high as 1,800°F (982°C), move from the furnace through the boiler tube section, where the contained water is heated to form saturated steam and dry steam. The flue gases continue through the economizer section to the air pollution-control device – such as an electrostatic precipitator, baghouse, or acid gas scrubber – where the flue gases are cleaned before being released into the atmosphere through a stack.

After the combustion process is completed, the grate system or rotary combustor gradually moves the waste onto the burnout grate where it is discharged into a wet or dry ash handling system that cools the residue and prevents dust from being created. The bottom ash produced from the combustion process in the furnace, and the fly ash or other materials produced in the air pollution-control device, are transported to landfills by truck or to a temporary onsite ash storage pit for later transport. The bottom and fly ash may be combined, or handled separately.

Mass-burn incineration produces ash residues amounting to 15 to 30 per cent by weight and 5 to 10 per cent by volume of the incoming municipal solid waste. Most facilities can produce an ash product that has less than 5 per cent combustible material and 0.2 per cent putrescible matter.

Recovery of ferrous and non-ferrous materials from the ash residue is possible in mass-burn systems. Many facilities have successfully utilized magnetic separators (with or without trommels) to recover ferrous material from the ash. Some systems have attempted to recover the remaining non-magnetic fraction in the ash, such as aluminum and glass, using various trommels, screens, jigs, and fluid separators.

3.4.2 Operations experience

Mass-burning incinerators have been used in Europe and Japan for municipal solid waste disposal for nearly 30 years, where their acceptance has been rapid and widespread. With over 500 facilities in operation worldwide in sizes ranging from 60 to 3,500 tons per day, mass-fired incineration is the most thoroughly demonstrated technology in the WTE field at this time.

This technology was introduced into the United States in 1967 at the US Naval Station in Norfolk, Virginia with the construction of a 360 ton per day waterwall plant to produce process energy for the Naval Shipyard. This plant was designed in America, using American machinery and equipment. Later plants were almost entirely designed using state-of-the-art European mass incineration technology.

The introduction of European technology into the United States has not been without difficulties and several of the earlier constructed plants encountered some mechanical problems. These highly reliable and rugged European systems were designed to burn solid waste that was somewhat different in composition to the American waste systems. Consequently, systems originally designed for European conditions required adjustments in the grate areas and furnace heat release rates in American plants. In addition, the higher chloride corrosion of the superheaters in American plants meant that designers needed to change the metallurgy of these boiler tubes, as well as limiting the upper stream pressures and temperatures to minimize tube corrosion. Scale-up problems also had to be overcome since many of the European units were designed for the 300 to 500 tons per day range. These problems have now been corrected, and most constructed mass-burn systems are still in operation today.

3.5 Modular combustion

A modular incinerator is a type of mass-burning, WTE unit prefabricated on a standardized modular basis in a factory. These plants operate on a starved air basis. Such units are shipped to the site in modules, ranging in design capacity from 10 tons per day to 200 tons per day, once they are installed. Several modules can be grouped together at a single location. These 'off the shelf' units can often be less costly to fabricate than the larger mass-burn facilities which require more costly field erection. Modular plants can also typically be constructed in some 15 to 20 months.

3.5.1 Process system

Modular incinerators are constructed in the United States with different process configurations. Some units are designed to incinerate solid waste under excess air conditions with either refractory furnaces or waste heat boilers or with waterwall boilers. However, the majority of units are designed to operate under starved air conditions with refractory furnaces and waste heat boilers (Figure 3.3).

A cross-section view of a typical modular combustion unit is illustrated in Figure 3.3. The majority of modular facilities have a tipping floor and utilize a front-end loader for simplicity in waste storage and feeding. Combusting takes place in either two or three stages. First, solid waste delivered to the facility is fed into the initial combustion chamber using a ram-type feeder. A moving ram slides back and forth over fixed steps within the chamber, causing the waste to tumble down one fixed section of the grate to the next fixed section. The waste is then transformed into a low-Btu gas, which is subsequently combusted in the secondary chamber, where auxiliary fuel is often fired under excess air conditions. A discharge ram on the back end of the combustion chamber feeds this incinerated waste into an ash quench bath.

The low-Btu gases produced by the combustion process in the first chamber are typically introduced into a secondary chamber where they burn at temperatures ranging from 980 to 2,000°F (527 to 1,093°C). Heat energy is recovered by convection in waste heat boilers in this secondary chamber, although waterwall boiler units for the primary and secondary chambers have been constructed.

Figure 3.3 Modular incinerators.
Source: Waste2Energy, Inc.

In recent years, several manufacturers have entered the modular plant marketplace using a batch oxidation process (BOS – Figure 3.4). The batch process integrates slow gasification and long exposure time at moderate temperatures followed by turbulent oxidation of gases at high temperature. After the waste is loaded into the primary chamber and sealed tight, an auxiliary burner is ignited to raise temperatures to about 392°F (200°C); the interior temperature is then monitored with controls, and maintained by allowing sub-stoichiometric amounts of air into the chamber during the gasification process. The combination of relatively low temperatures and only sub-stoichiometric amounts of air in the primary chamber during gasification does not disturb the gasification bed, which is said to minimize particulate emissions, heavy metals, and many combustion gases. Depending on the waste type and system layout, the waste reduction process in the primary chamber will take approximately 10 to 15 hours.

Emissions produced during the gasification process pass through to the preheated secondary chamber – also called an 'afterburner' – where most of the remaining air emissions are eliminated. As the gases from the primary chamber enter a preheated

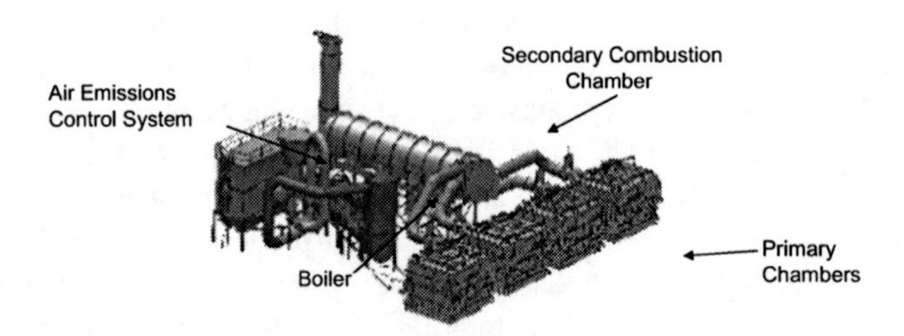

Figure 3.4 Cross-section of batch oxidation system, modular facility.
Source: Waste2Energy, Inc.

secondary chamber, auxiliary burners and excess oxygen create a very turbulent high temperature environment (typically between 1,562 and 2,192°F [850 and 1,200°C]). For most applications within the European Union (EU), 1,562°F (850°C) is the required minimum, though 2,012°F (1,100°C) is required for halogenated wastes, and in North America, 1,800°F (982°C) is usually required. Additionally, residence time in the secondary chamber is important for proper destruction of emissions from the primary chamber. In both the EU and North America, a minimum residence time of two seconds is required.

These modular technologies, while being cheaper, provide a burn out that oftentimes is not as good as mass-burn. Also, energy recovery is lower because the size of the boiler is quite small in comparison to mass-burn heat surface ratios. Life expectancy of such a plant is anticipated to be around 10 to 15 years as opposed 30 years for mass-burn.

3.5.2 Operating facilities

There have been many more modular WTE incinerators constructed in the United States than either the mass-burn or refuse-derived fuel systems. In 1977, the first modular incinerator began operations in North Little Rock, Arkansas to produce steam for Koppers Industries' Forest Products Division. Since then, some 50 modular systems have been built in the United States (Table 3.3), almost exclusively to

Table 3.3 Comparison of active modular combustion facilities

Location	Startup	Design capacity (tons/day)	Energy generation	Capital cost (US$ millions)
Auburn, ME	1992	200	Steam	4.0
Joppa, MD	1988	360	Steam	10.0
Pittsfield, MA	1981	360	Steam	10.8
Alexandria, MN	1987	80	Steam/Electricity (0.5 MW)	4.2
Fosston. MN	1988	80	Steam	4.5
Perham, MN	1986/2002	116	Steam/Electricity (2.5 MW)	6.0
Red Wing, MN	1982	90	Steam	2.5
Fulton, NY	1985	200	Steam/Electricity (4 MW)	14.5
Almena, WI	1986	100	Steam/Electricity (0.27 MW)	2.7
Husavik Municipality, Iceland	2006	20	Steam	3.5
Scotget, Scotland	2009	180	Electricity	40.0
Turks and Caicos Is.	2008	4	None	1.0
US Air Force, Wake Island	2009	1.5	None	0.5
US Department of Defense, Kwajalein Atoll	2007	32	None	5.0

produce process steam for neighboring industries. Some of these systems, for example, the plant in Collegeville, Minnesota, have utilized the community's solid waste as a fuel to produce steam to a district heating loop during the winter, and electricity during the summer. Many of the newer facilities have incorporated electricity production capability.

Modular combustion units offer a lower capital cost and simplicity to communities considering WTE systems than with the larger field-erected mass-burning systems. These systems are generally reliable and backed by many years of successful operating experience. The newer BOS systems appear to offer substantially lower costs of operations and maintenance. For example, the manpower required to operate these systems is generally minimal, with one worker required to load the primary chamber and discharge the ash stream within an hour. Many suppliers claim nearly complete burn out between energy recovery and recycling. The ash remaining is reported to be about 3 to 8 per cent of the original volume (depending on waste composition). Lastly, these systems are modular and can be used or decreased in size easily.

3.6 Refuse derived fuel (RDF) systems

Several American corporations have developed technologies that pre-process solid waste to varying degrees to separate the non-combustibles from the waste stream. By undergoing processing steps of hammering, shredding, or hydropulping, the combustible fraction of the waste is transformed into a fuel, which can then be fired in a boiler unit specifically dedicated for this type of refuse-derived fuel, or co-fired with another fuel, such as coal, shredded tires, or wood chips. The fuel produced can thus be utilized in equipment that can have higher efficiencies than mass-fired units resulting in greater electricity or steam output. However, the front end processing of the solid waste into a fuel has been one of the problem areas of this type of refuse disposal technology.

3.6.1 Processing systems

The processing of solid waste into a refuse-derived fuel uses both wet and dry processing systems.

3.6.1.1 Wet RDF processing

Wet RDF processing utilizes hydropulping technology adapted from the pulp and paper industry. The solid waste is fed into a large pulper, which acts very much like a kitchen blender, where it is mixed with water forming a slurry. The resulting slurry is transferred to liquid cyclones that separate the combustible from the non-combustible fractions. The combustible fraction is then mechanically dewatered and dried to a moisture content of 50 per cent solids before being introduced as fuel into a dedicated boiler. However, the energy efficiency of this process is reduced by the need for drying.

This wet processing system was first utilized in a 150 ton per day pilot plant in Franklin, Ohio from 1972 to 1979. This plant has now closed. The first full-scale 2,000 tons per day system was constructed by Black Clawson, a subsidiary of Parsons and Whittemore Company, at Hempstead, New York. This plant experienced several early operational difficulties, resulting in its closure. A 3,000 tons per day sister facility is currently in operation in Dade County, Florida. However, the operators of this facility abandoned the hydropulping equipment.

3.6.1.2 Dry processing systems

Since the early 1970s, there have been several dozen facilities constructed in the United States that process solid waste into a refuse-derived fuel through dry processing systems. Such dry processing systems are classified according to the type of products that can be produced: fluff RDF, densified RDF, and powdered RDF. A cross-section of a typical RDF system is illustrated in Figure 3.5.

Given the number of potential products, the type of specific technology used to process the solid waste into a refuse-derived fuel can vary from one location to another. Typically, however, solid waste delivered to an RDF facility is unloaded onto the receiving floor, or at some locations into a refuse storage pit. The waste is then transported with the use of a feed conveyor system to a size reduction unit, which reduces the particle size of the waste. At many facilities, machinery such as flail mills, trommels, and magnetic separators are used to pre-sort the waste prior to its being fed into a hammermill or shredder for size reduction. Depending on the particular type of RDF fuel required, further processing equipment is utilized after shredding, such as air classifiers, densifiers, and trammel screens. The result of the dry processing system is a refuse-derived fuel, which can be combusted either in existing utility boilers or in boilers specifically designed for the type of RDF produced (dedicated boilers).

For many years, the plan of burning RDF in existing electric utility boilers seemed an obvious solution for communities that needed a good way to dispose of garbage. It was hoped that existing boilers and air pollution control equipment could be utilized, thus saving these communities considerable capital expense. However, since the early 1970s, RDF has been tried as a supplemental utility fuel with mixed success. Since 1970, utilities in the United States have co-fired RDF in their system boilers; only three are still burning RDF.

Shortly after the beginning of the first demonstration project in St. Louis in 1972, it became apparent that burning RDF in utility boilers resulted in a lowering of their normal efficiency and reliability. When RDF was fired in the high temperature, utility boilers, the non-combustible materials in solid waste, such as glass and metals, melted into slag that fouled the boiler tubes, heat exchangers, and furnace walls. Burning of the plastic compounds in the solid waste, which released chlorine, also resulted in increased corrosion of boiler parts. In addition to these problems, ash handling, air pollution control, and materials handling systems soon became overloaded and were subject to frequent outages. In short, what had seemed to be a good way to dispose of solid waste resulted in an unexpected headache for utility companies. The initial

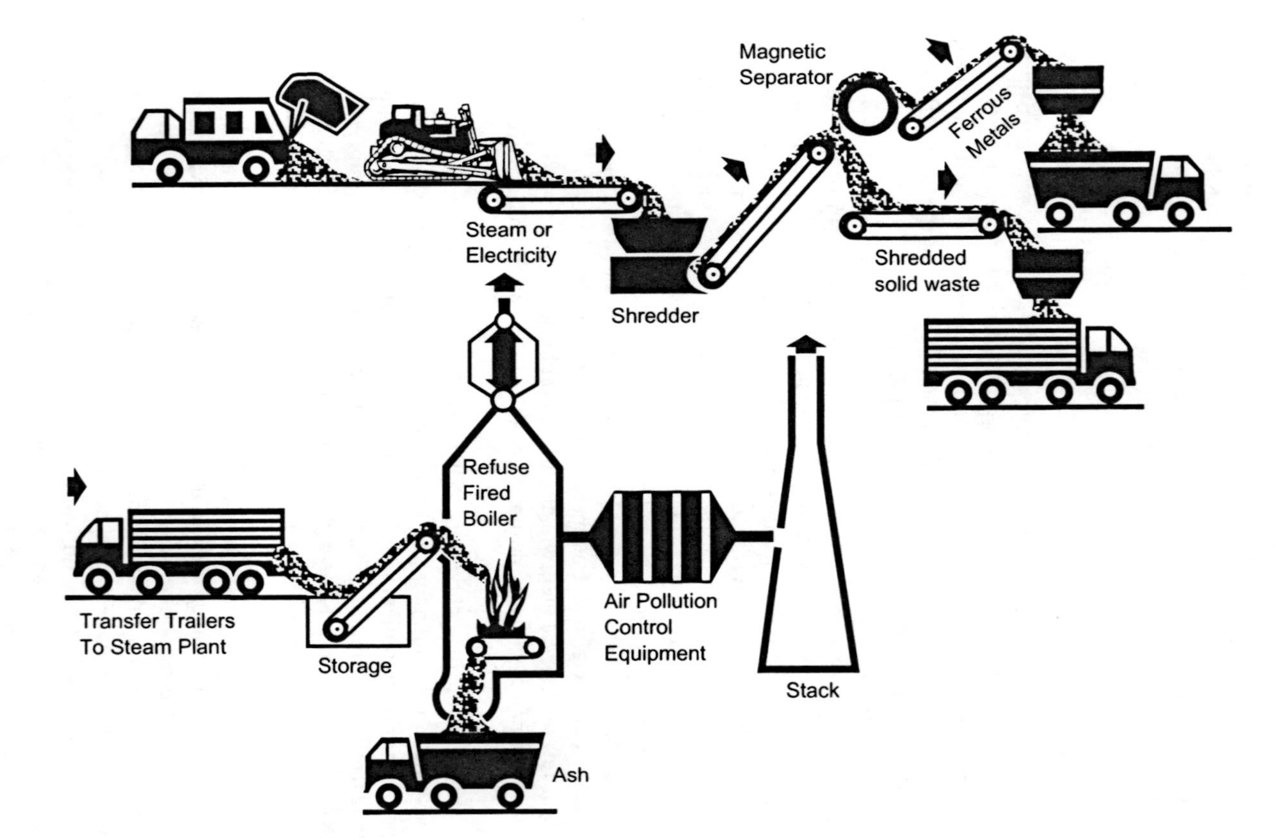

Figure 3.5 Cross-section of typical RDF system.

optimism shown for this technological fix to solve an emerging garbage disposal problem has consequently not been realized.

The recent emphasis on burning RDF has focused on systems using 'dedicated' industrial-type boilers. The term 'dedicated' refers to a boiler system specifically designed and constructed to burn RDF as its primary, not supplemental, fuel. There is a variety of different types of technologies used in such boilers: suspension-fired; semi-suspension fired (spreader stoker); pyrolysis; and fluidized bed.

The semi-suspension fired or spreader-stoker, furnace boiler is perhaps today's most commonly utilized technology. Spreader-stoker technology has been utilized successfully for decades for incineration of a variety of different solid fuels. With this system, RDF, previously prepared to specific size characteristics, is introduced at a controlled rate to pneumatic RDF distributors located at the front wall of the furnace. High-pressure air is delivered to these distributors to assure that the RDF is fed evenly.

The RDF, so introduced, ignites over the grate area and burns partially in suspension. Materials, which are left unburned, fall to the traveling stoker where they are combusted before the ash is discharged.

The experience with these RDF systems has varied. It appears that most installations have had problems with RDF feeder equipment, resulting in extensive retrofits and technical modifications. However, where high quality RDF has been introduced with most metals and glass removed, the RDF burning experience has generally been good. On the other hand, the experience of co-firing of RDF with coal has been generally poor. The combination of coal and RDF appears to increase problems with ash clinker and slagging and wear of the lower furnace walls.

In summary, RDF with waste shredding was supposed to be a smart move to be able to burn garbage in a 'cheaper' biomass or coal designed boiler. However, the cost of shredding and the safety risk involved is not optimal. Feeding is a major headache and the 'biomass' boiler with only one combustion empty path is more costly, is generally cost prohibitive to maintain, and requires a huge site footprint.

In recent years, a 'second generation' RDF facility design has evolved. This combines a simple front-end processing unit with a smaller mass burn unit on the back. The front-end processing unit does not shred the waste, but is used only to sort, recycle, open bags, and increase waste LHV. Because of the recycling upfront, the mass-burn unit size is smaller than for a conventional mass-burn and more energy is recovered. This is the new wave of WTE plant, commonly called an 'advanced thermal recycling energy from waste facility', which involves recovering biogas from organic waste as well.

3.7 Fluidized bed systems

Fluidized-bed technology has been in existence for well over 50 years, principally in Europe. This technology has been utilized for burning a variety of low-quality fuels such as: high-sulfur coal; peat; tires; sludge; waste oils; and biomass wastes. Within the last decade, fluidized-bed technology has undergone extensive refinement.

Three types of systems have been developed for burning coal: the bubbling-bed; the dual-bed; and the circulating-bed. For combusting solid waste, the circulating-bed process has proven to be the best method. Utilizing this system, refuse-derived fuel is injected into a fluidized-bed combustor composed of steel and refractory brick liner and a fluidized bed of limestone and sand. The heated bed of thermally inert materials behaves like a fluid because high-velocity air is injected into the combustion chamber. The RDF is introduced into, or on top of, this circulating-bed where it is combusted at temperatures typically in the range of 1,550–1,650°F (843–899°C). Energy from the flue gases is removed using a waste heat boiler, heat exchanger tubes, in-bed tubes, or waterwalls.

Although fluidized-bed technology has been utilized in the utility industry, there is little demonstrated experience in the United States to date with municipal solid waste. European experience in fluidized-bed incineration of solid waste comes from several plants, principally in Sweden, where RDF is incinerated in conjunction with other fuels. Limited operational data are available because these facilities have only begun commercial operation within the last few years.

In the United States, fluidized-bed incineration of RDF has been tested at a number of pilot locations in Menlo Park, California; at West Virginia University; in Franklin, Ohio; and at French Island, Wisconsin. A commercial-scale system for co-firing refuse and sludge was constructed in Duluth, Minnesota.

The experiences at these pilot projects, and in Duluth, indicate that the use of fluidized-beds to incinerate municipal solid waste is in a research and development phase. Success at these projects has been shown to depend on maintaining a stable bed of limestone, sand, and RDF. Unfortunately, experience has shown that the ash and residue from the RDF fuel can change the physical and chemical composition of this bed, necessitating continual replacement of the bed. A further problem is non-combustible particles in the RDF, such as glass and metals, which can accumulate in the bed and fuse with the ash. The result is that the bed becomes de-fluidized.

To overcome these problems, extensive front end processing of the RDF fuel is necessary to prepare 'a perfect fuel'. Alternatively, only selected waste sources (paper and plastic) are used as feedstocks for the plant. The degree of the processing required could negatively impact the operational economics of such systems. Several Japanese plants have been successful in operating fluidized beds by investing hugely on front end process, but the tip fees are very expensive (US$ 200/ton tip fees). However, such systems have shown that acid gases can be removed with considerable efficiency. Thus, expenditures for air pollution control equipment could be expected to be significantly reduced.

3.8 Emerging waste conversion technologies

3.8.1 Summary of technologies

Conversion technologies include an array of emerging technologies that are capable of converting post-recycling residual solid waste into useful products and chemicals, including ethanol and biodiesel, and clean renewable energy. The technologies may be thermal, chemical, or biological. These technologies have been used successfully to manage MSW in Europe and Asia, but, at the time of writing, commercial development in the United States is still in the design stage.

Technologies that appear amenable for converting organic and other materials into energy, ethanol and other products, include hydrolysis, gasification, anaerobic digestion, and plasma arc. The following sections briefly describe these technologies; Table 3.4 provides a very general comparative overview of these technologies. Throughout this section, we use the terms 'conversion technologies' and 'alternative technologies' interchangeably to describe technologies that are being considered for MSW processing and conversion to energy and other products.

3.8.1.1 Hydrolysis

Hydrolysis is a chemical decomposition process that uses water to split chemical bonds of substances. There are two types of hydrolysis: acid and enzymatic. Feedstock that may be appropriate for acid or enzymatic hydrolysis is typically plant-based materials containing cellulose. These include forest material and sawmill residue, agricultural residue, urban waste, and waste paper.

Ethanol facilities could be co-located at MRFs where existing materials are already collected and the existing solid waste transportation infrastructure could be utilized. Ethanol facilities co-located at MRFs could take advantage of the existing solid waste collection and transportation infrastructure. Figure 3.6 includes a typical hydrolysis process.

3.8.1.2 Gasification

Gasification is a process that uses heat, pressure, and steam to convert materials directly into a gas composed primarily of carbon monoxide and hydrogen. Gasification technologies differ in many aspects but rely on four key engineering factors:

- Gasification reactor atmosphere (level of oxygen or air content);
- Reactor design;

Table 3.4 General overview of conversion technologies

Technology	Amenable feedstock	Feedstock requirements	Emissions/Residues
Acid or Enzyme Hydrolysis	Cellulosic material	Cellulosic feedstock	Wastewater, CO_2
Gasification	Biomass, MSW	Drier feedstock, high carbon	Ammonia, NO_x, tars, oil
Anaerobic Digestion	Manure, Biosolids	Wet material, High nitrogen	Wastewater, CH_4, CO_2, H_2S
Plasma Arc	MSW	Removal of metals and inerts	Slag, scrubber water

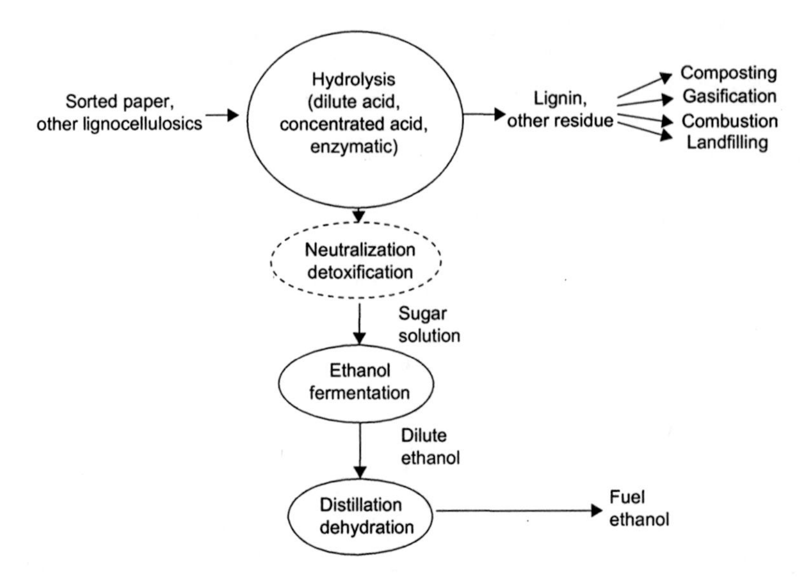

Figure 3.6 Typical hydrolysis process.
Source: SCS Engineers.

- Internal and external heating;
- Operating temperature.

Typical raw materials used in gasification are coal, petroleum-based materials, and organic materials. The feedstock is prepared and fed, in either dry or slurried form, into a sealed reactor chamber called a gasifier. The feedstock is subjected to high heat, pressure, and either an oxygen-rich or an oxygen-starved environment within the gasifier. Most commercial gasification technologies do not use oxygen. All require an energy source to generate heat and begin processing.

There are three primary products from gasification:

- Hydrocarbon gases (also called syngas);
- Hydrocarbon liquids (oils);
- Char (carbon black and ash).

Syngas is primarily carbon monoxide and hydrogen (more than 85 per cent by volume) and smaller quantities of carbon dioxide and methane. Syngas can be used as a fuel to generate electricity or steam, or as a basic chemical building block for a multitude of uses. When mixed with air, syngas can be used in gasoline or diesel engines with few modifications to the engine.

As in the case of ethanol conversion facilities, gasification facilities could be co-located at MRFs to take advantage of the current solid waste transportation infrastructure. In addition, co-location at MRFs would ensure that recyclable materials are removed beforehand and only residuals would be sent to a gasifier. If a gasification facility is co-located at a landfill that accepts MRF residuals, the gasification facility could utilize landfill gas in the gasification process or could work in tandem with a landfill gas-to-electricity project. Figure 3.7 shows a typical gasification system.

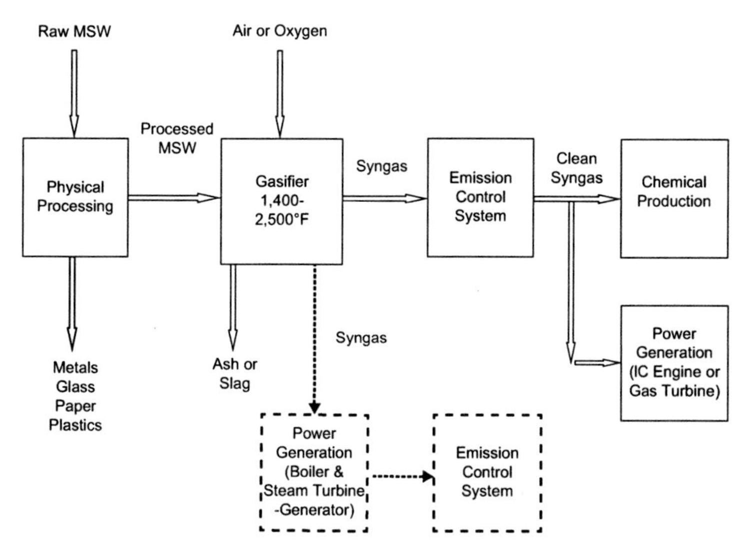

Figure 3.7 Typical gasification system for power generation or chemical production.
Source: SCS Engineers.

3.8.1.3 *Anaerobic digestion*

Anaerobic digestion is the bacterial breakdown of organic materials in the absence of oxygen. This biological process generates a gas, sometimes called biogas, principally composed of methane and carbon dioxide. This gas is produced from feedstock such as biosolids, livestock manure, and wet organic materials.

The anaerobic digestion process occurs in three steps:

- Decomposition of plant or animal matter by bacteria into molecules such as sugar;
- Conversion of decomposed matter to organic acids;
- Organic acid conversion to methane gas.

Anaerobic processes can occur naturally or in a controlled environment such as a biogas plant. In controlled environments, organic materials such as biosolids and other relatively wet organic materials, along with various types of bacteria, are put in an airtight container called a digester where the process occurs. Depending on the waste feedstock and the system design, biogas is typically 55 to 75 per cent pure methane. A typical anaerobic digestion process system is shown in Figure 3.8.

3.8.1.4 *Plasma arc*

Plasma arc technology is a non-incineration thermal process that uses extremely high temperatures in an oxygen-starved environment to completely decompose waste into very simple molecules. Plasma arc technology has been used for many years for metals processing. The heat source is a plasma arc torch, a device that produces a very high temperature plasma gas. A plasma gas is the hottest sustainable heat source available, with temperatures ranging from 2,700 to 12,000°F (1,482 to 6,649°C).

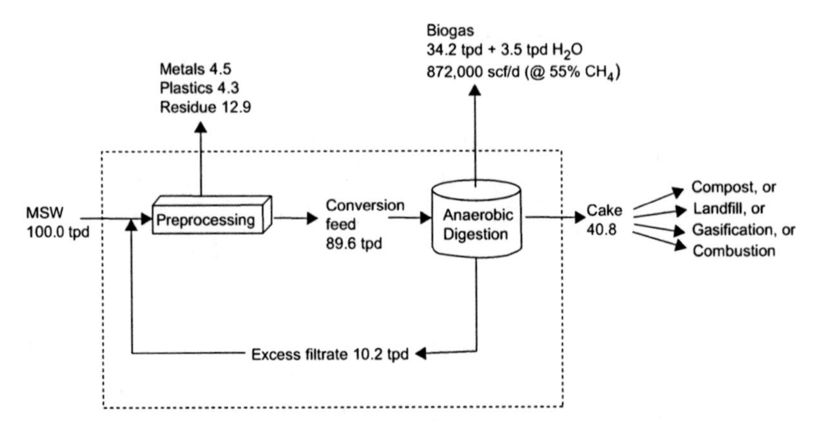

Figure 3.8 Typical MSW anaerobic digestion process system.
Source: SCS Engineers.

A plasma arc system is designed specifically for the type, size, and quantity of waste material to be processed. The very high temperature profile of the plasma gas provides an optimal processing zone within the reactor vessel through which all input material is forced to pass. The reactor vessel operates at atmospheric pressure.

The feedstock can be almost completely gasified, while non-combustible material, including glass and metal, is reduced to an inert slag. The product gas typically has a heating value approximately one-quarter to one-third the heating value of natural gas (natural gas has a value of approximately 1,040 Btu/standard cubic foot); therefore, it may be used as an efficient fuel source for industrial processes, including the generation of electricity, and the production of methanol and ethanol. The slag can be used in the construction industry or for road paving. All other byproducts, such as scrubber water

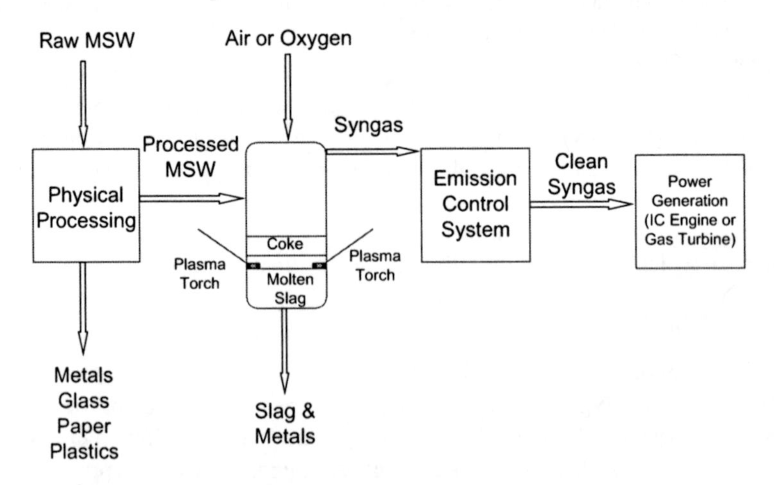

Figure 3.9 Typical plasma gasification system.
Source: SCS Engineers.

and cyclone catch material, can be recycled into the process for reprocessing to alleviate disposal requirements. A typical plasma gasification system is shown in Figure 3.9.

3.9 Summary

As noted in the preceding sections, there is a variety of WTE technologies that have been employed worldwide over the past few decades, to process and dispose of municipal solid waste. As communities undertake the strategic planning process for implementation of the proposed WTE facility, there are a number of issues that should be considered, including the strengths and weaknesses of each of these technologies, current operating history of each particular technology, and overall estimated capital and operational costs.

Table 3.5 provides an example of a summary of major WTE technologies, and shows how they are applied, based on the authors' long-term knowledge of the WTE industry. We are of the opinion that technologies that may be particularly of interest to relatively small Third World locations are modular starved air and gasification. Both these technologies have been proved to be technically successful in the processing of the quantities of solid waste, which are currently generated in these communities. As noted, the other available WTE technologies are either currently unproven or typically implemented at a design processing capacity more than a magnitude greater than the size range needed in these communities. In these cases, the initial capital and operating costs would be substantially greater than afforded by implementation of a facility using modular starved air or gasification WTE technologies.

Table 3.5 Evaluation of WTE technologies relative to RMI's solid waste generation levels

	Applicable to RMI		
Technologies	**Yes**	**No**	**Comment**
Mass-Burn		X	Commercially successful worldwide, but has high capital and operating costs for RMI's waste generation levels
RDF		X	Commercially successful worldwide, but has high capital and operating costs for RMI's waste generation levels
Modular Starved Air	X		Excellent operational experience for RMI's waste generation levels
Gasification	X		Excellent operational experience for RMI's waste generation levels; limited ash disposal requirements
Plasma Arc Gasification		X	Limited current operating experience; high capital and operating costs
Biochemical		X	Not yet commercially proven
Anaerobic Digestion		X	Limited on-island markets for compost

4 Solid waste composition and quantities

Chapter Outline

4.1 Introduction

Proper planning of a WTE facility requires that a reliable database on solid waste characteristics and quantities currently generated and expected to be generated within the service area of the facility, is available [1]. Such data are necessary, not only for determining the current refuse disposal needs of the community, but also to determine the overall future requirements of the solid waste disposal system. The quantities of solid waste generated by the community may have an impact on the initial sizing of proposed WTE facilities, emergency/ash residue landfills, and other ancillary facilities, as well as determine the most efficient location of such facilities, so as to minimize transportation costs.

Further, the quantities of solid waste generated by the community and available for processing by the facility are also an important factor in determining the financial

Waste-to-Energy. DOI: 10.1016/B978-1-4377-7871-7.10004-8

feasibility of a proposed facility since the revenues generated through the sale of energy and recovered materials directly correlate to the amount of solid waste received by such facilities. The composition of a community's waste is also a critical factor since it can affect the energy content of the waste received by a facility, as well as the quantities of recyclable materials and residues that may be generated. Thus, waste composition can influence the design criteria and economics of any WTE facility. Recycling programs for inert materials and metals such as glass, aluminum, and ferrous are complementary to WTE systems.

4.2 Types of solid waste

For the purposes of our discussion, municipal solid waste (MSW) typically refers to solid or semi-solid discarded materials resulting from industrial, commercial, agricultural, institutional, and residential operations, but does not include solids or dissolved materials in industrial or domestic sewage (waste sludge), or hazardous wastes. The latter wastes are now not permitted to be disposed of at solid waste facilities by most regulatory agencies because of their potential impacts on the environment. These materials may include volatile, chemical, biological, explosive, flammable, and radioactive materials. While solid waste from most communities contains small amounts of such similar materials as solvents, paints, and household cleaners, they constitute a minimal potential impact upon the environment since they are a relatively small fraction of the total volume of waste generated by most communities. Consequently, regulatory agencies in the United States and Europe do not currently consider municipal solid waste as hazardous.

The following classifications of municipal solid waste materials will be used as a basis for subsequent discussion:

- Residential waste – Household materials generated by residents in single-family and multifamily dwellings. These mixed household wastes include kitchen and food wastes (often called garbage). They are highly putrescible, will decompose rapidly, and pose a health and safety problem if left uncollected. Solid waste generated by households also includes combustible materials such as paper, cardboard, plastics, and garden trimmings, and noncombustible materials such as glass, metals, and soil.
- Commercial waste – Generated by wholesale and retail stores, restaurants, markets, office buildings, hotels, motels and other similar establishments, and large institutional facilities such as hospitals, prisons, schools, and religious institutions. The wastes generated by the commercial sector are very similar in physical characteristics and composition to that generated by residential units.
- Industrial wastes – Solid waste generated by various types of manufacturing and industrial operations, excluding such hazardous materials as solvents, oils, chemicals, and by similar manufacturing establishments.
- Special wastes – Because of their physical characteristics, these solid wastes require special or extraordinary handling. This includes bulky materials such as abandoned vehicles, used tires, white goods (e.g., refrigerators), furniture, and materials generated from demolition and construction projects such as soil, stones, concrete rubble, bricks, lumber, and shingles.

4.3 Solid waste quantities

The quantities of solid waste generated by a community can be determined from a number of data sources. The primary sources of data are records kept by landfill operators in the community over several years. These data, however, may be incomplete due to poor or variable recordkeeping from one location to another. In addition, it may be impossible to determine accurately the full waste load of certain waste collection routes in a community. Furthermore, the waste disposed of at such locations may be recorded on a volume basis rather than by being weighed on a scale. Additionally, waste may also be disposed of illegally, never entering the community's waste disposal system. These problems are not uncommon throughout the United States, and estimating procedures must be utilized to develop reasonable solid waste generation rates for the community.

Population data are useful to help compare existing records of solid waste tonnages for a community. A per capita solid waste rate can be developed, which can then be compared to generation rates for other communities with similar characteristics to determine reasonableness of such a solid waste generation rate. In some instances, it may be useful to develop per capita generation rates for different solid waste classifications (e.g., residential, commercial, and industrial) to assist in this comparison and to determine the reasons, if any, for differences between the community's per capita generation rate and other similar communities'. In this way, high or low rates may be easily explained due to certain specific local factors. The key point of such an analysis is to determine a reasonable solid waste generation rate that the community can guarantee to deliver to a WTE facility, and upon which long-term waste projections can be based. In this way, the WTE facility can be designed prudently, and sized to accommodate future expansion due to population growth.

4.3.1 Conducting an MSW weighing program

In order to determine the amount of waste available for disposal in a WTE facility, weigh data are typically collected from landfills and transfer stations in the community's solid waste system.

Where comprehensive scale data are unavailable, the feasibility analysis for the WTE project must, therefore, rely on short-term weighing programs to collect reliable data to size the facility. Such programs are typically implemented in developing areas, as illustrated by a study conducted by the authors in American Samoa.

The feasibility analysis in this study relied on the results of a three-week weighing program, conducted by the American Samoa Power Authority (ASPA) during three six-day periods, the last one concurrent with the consultant's sorting activities for a waste composition study. Since the ASPA landfill did not have, at the time, an operational scale, all haulers, residents, and businesses were asked to weigh in and out at a scrap metal yard during this period. With some modifications, the raw data were converted to approximate the waste stream that would be typically disposed of at the

ASPA landfill or proposed WTE plant. Tables 4.1 and 4.2 summarize the results of the weighing program, as well as estimates for annual disposal.

For analysis and planning purposes, MSW disposal estimates were divided into substreams. For this study, substreams were defined by combining the type of generator that created the waste and the type of truck that was used to deliver material to the site. A total of 10 substream categories were identified:

1. ASG/ASPA pickup truck – Includes waste collected by government officials and delivered in pickup trucks. Also, includes sources such as EPA (Environmental Protection Agency) and the Parks and Recreation Department.
2. ASCC pickup truck – Includes waste collected by government officials and delivered in pickup trucks from AS Community College.
3. ASPA bucket truck – Includes waste collected by government officials and delivered in bucket trucks.
4. ASPA flat bed truck – Includes waste collected by ASPA and delivered to the landfill in flat bed trucks.
5. ASPA side loader – Includes waste collected by ASPA and delivered to the landfill in side loader trucks.
6. Family trash – Waste hauled by individuals and delivered in pickup trucks.
7. Starkist (reg. trash) – Mixed waste hauled by Starkist directly to the landfill.
8. Starkist (waste fish) – Whole fish hauled by Starkist directly to the landfill.
9. Commercial front end loader – Includes waste collected by ASPA, Paramount, and T&T from businesses, institutions, industries, and residential units.
10. Mixed commercial – Includes waste hauled directly by businesses, institutions, and industries in flat bed trucks, pickup trucks, or bucket trucks.

Table 4.1 Results of three-week weighing program

Account/Name, truck type	Sector	Net weight (T)	% of Sector	% of Total
ASG/ASPA Pickup Truck	ASG & Res.	18	5.5	
ASCC Pickup Truck	ASG & Res.	1	0.3	
ASPA Bucket Truck	ASG & Res.	1	0.2	
ASPA Flat Bed Truck	ASG & Res.	123	36.7	
ASPA Side Loader	ASG & Res.	163	48.6	
Family Trash	ASG & Res.	27	8.1	
Total, ASG & Residential		**336**		**27.7**
Three Week Average, ASG & Residential		**42**		
Starkist (Reg. Trash)	Commercial	84	9.6	
Starkist (Waste Fish)	Commercial	54	6.1	
Commercial Front End Loader	Commercial	683	78.0	
Mixed Commercial	Commercial	55	6.3	
Total, Commercial		**876**		**72.3**
Three Week Average, Commercial		**125**		
Total, Three Weeks		**1,211**		

Source: Reference [2].

Table 4.2 Estimates of annual disposal, based on three-week weighing program

Account/Name, truck type	Sector	Net weight (T)	% of Sector	% of Total
ASG/ASPA Pickup Truck	ASG & Res.	318	5.5	5.5
ASCC Pickup Truck	ASG & Res.	17	0.3	0.3
ASPA Bucket Truck	ASG & Res.	12	0.2	0.2
ASPA Flat Bed Truck	ASG & Res.	2,136	36.7	36.7
ASPA Side Loader	ASG & Res.	2,829	48.6	48.6
Family Trash	ASG & Res.	473	8.1	8.1
Total for Year, ASG & Residential		**5,785**		**27.7**
Starkist (Reg. Trash)	Commercial	1,457	9.6	6.9
Starkist (Waste Fish)	Commercial	929	6.1	4.4
Commercial Front End Loader	Commercial	11,832	78.0	56.4
Mixed Commercial	Commercial	957	6.3	4.6
Total for Year, Commercial		**14,246**		**72.3**
Total for Year		**20,960**		

Source: Reference [2].

Three additional substreams (ATSCA Boom Truck, ASPA Flatbed, and Pacific Island Cable) were not included in the waste characterization portion of the study because no samples were taken from these very small substreams. However, these totals are included under the 'Mixed Residue' calculation of the estimated heating value of the waste stream. These three substreams totaled 36 tons; therefore, a total of 20,996 tons was used in the heating value section of this report only.

In addition to estimates from above, this study included estimates for waste oil and tire generation as potential sources for the WTE facility.

4.3.2 *Waste oil*

Waste oil that is currently collected and transferred to the canneries for processing and recycling, could be incinerated at the WTE facility to minimize disposal expenses and to add a valuable fuel source (Figure 4.1). It is estimated that 120,000 gallons (510 tons) of oil will be available for use every year.

4.3.3 *Tires*

A site trip to the scrap yard suggested that several thousand tires are available as feedstock to the proposed WTE (Figure 4.2). Using a conversion factor of 220 lbs/cy, it was estimated that there are approximately 20 tons of tires available for incineration or recycling off island.

A recent survey of metals markets on Samoa estimated that auto dealerships on the island recorded sales of 425 new light vehicles per year. Further, the territory government reported a total of 875 new vehicle registrations, indicating that a

Figure 4.1 Waste oil holding tanks at ASPA power plant.

substantial number of vehicles are imported onto the island, either by individuals or other small companies. The survey also provided some details on the number of commercial vehicles (e.g., trucks, vans, and buses). Respondents indicated that they had brought in a total of 38 commercial vehicles in 2006, mostly large trucks and buses. Using an average tire generation rate of one tire per person year, it is estimated that there would be approximately 66,000 tires available per year for the proposed WTE facility. Using a conversion factor of 36 pounds per tire, it was estimated that there would be approximately 1,200 tons of tires per year available for thermal treatment.

4.3.4 Estimated waste stream available for a WTE facility

Typical WTE facilities are run continuously on a 24-hour, seven days a week operational pattern. To estimate the island's waste stream available for this proposed WTE facility, the data from the weighing program were converted into a seven-day average for plant sizing purposes. As shown in Figure 4.3, the weigh data appeared to show that approximately 62 tons per day of MSW, oil, and tires would be available as feedstock for the plant. For feasibility analysis, various assumptions for plant sizing were made based on future population and waste stream growth and average annual downtime of different types of WTE facility technologies. These daily facility sizing needs were then projected over a ten-year planning horizon (Table 4.3).

Figure 4.2 Tire pile at scrap metal yard.

4.4 Waste composition methodology

This section briefly summarizes sorting methodology used for determining the composition of the waste stream proposed for a WTE facility. For each sample selected, the following data were collected:

- Unique sample identification number;
- Date the sample was taken;
- Waste sector (residential, commercial, industrial, institutional, or mixed);
- Type of collection (house to house, bin, mixed, or self-haul);
- Name of the hauler, if applicable;
- Size of the entire load (in cubic yards); and
- Site conditions that may impact the results of the study.

The Sampling Manager directed the driver to tip the load in an elongated pile near the operating face of the landfill. The pile was then divided into an imaginary eight-cell grid, as shown in Figure 4.4, and a sample of waste weighing approximately 200 pounds was extracted from a randomly selected cell using a loader operated by facility staff. This material was then transported to the sorting area of the landfill for weighing and sorting. For material delivered by flatbed truck, the Sampling Manager similarly divided the load into a grid and randomly selected a portion for sampling.

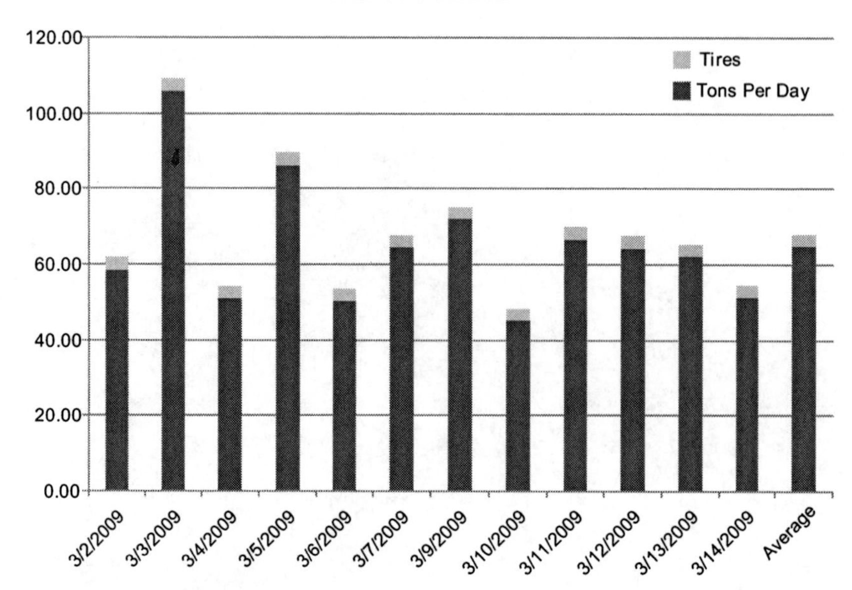

Figure 4.3 Estimated waste stream available for the proposed WTE facility on American Samoa.
Source: Reference [2].

Table 4.3 WTE plant sizing assumptions

	Year										
Assumptions	**0**	**1**	**2**	**3**	**4**	**5**	**6**	**7**	**8**	**9**	**10**
Waste Growth (1.5%/Year)	62	63	64	65	66	67	68	69	70	71	72
WTE Downtime @10%	69	70	71	72	73	74	75	76	78	79	80
WTE Downtime @15%	73	74	75	76	77	79	80	81	82	83	85

Source: Reference [2].

4.5 Waste sorting

4.5.1 Hand-sort procedure

The sorting team hand-sorted approximately six to eight samples daily. Samples can be sorted into 53 material categories (as defined in the Program Design), and the material in each category weighed. Material that was not defined as belonging to the first 52 high-priority material categories was included in the 53rd material category 'mixed residue'. For each sample, the Sorting Manager reviewed the sorted material

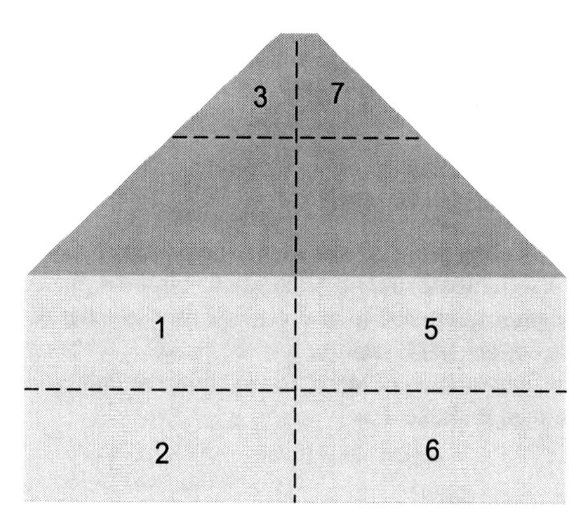

Figure 4.4 Visual overlay showing 'cells' of material.
Source: SCS Engineers.

for homogeneity before the sample components were weighed. The weight for each sorted material category was then recorded on a sampling form.

The sample was placed on the sorting tables and sorted into the pre-determined material categories by hand. Separated materials were placed in containers (laundry baskets) to be weighed and recorded. Members of the sorting crew were assigned material categories to focus on. The sorting crew then sorted the waste samples until no more than a small amount of homogeneous fine material ('mixed residue') remained. This was determined by the Sorting Manager. The overall goal was to sort each sample directly into the material categories in order to reduce the amount of indistinguishable fines or miscellaneous categories.

4.5.2 Visual characterization procedure

The Sampling Manager randomly selected loads to visually characterize. The method used is particularly useful for identifying recoverable materials that may be present in large quantities, and characterizing waste loads that contain bulky items and waste streams that tend to have substantial composition variation within individual loads (for example, loads that are half dirt and half lumber, separated at opposite ends of a truck).

The first step in visually estimating the composition of a selected load was to measure the volume of the waste. The visual estimator then recorded the estimated percentage of the load corresponding to each material class, and thereafter the estimated percentages for specific material categories within the material classes.

4.5.3 Disposition of sampled waste

Once all material types found had been weighed and documented for each sample, the material was delivered back to the working face of the landfill for disposition.

4.5.4 Testing of samples

Samples were collected during each season and sent to a laboratory for analysis of moisture content.

4.5.5 Waste composition analysis

Data from the waste sorting program was used to help prepare an overall composition of the community's solid waste stream. As shown in Table 4.4, the results of the short-term weighing program were used to calculate the total number of tons available as potential feedstock for the WTE facility.

Examples of the composition of MSW from China, the Philippines, and European countries are presented in Table 4.5.

4.5.6 Heating value (Btu)

Since the sale of energy plays an important role in the economic feasibility of a WTE project, the heating value of the waste stream is a key design factor. The heating value of solid waste is measured by Btu (British thermal units) per pound (or KJ/kg) of refuse. One Btu is defined as the amount of energy required to raise the temperature of one pound of water one degree Fahrenheit. Thus, the heating value is a basic measure of the heat energy released through the incineration of solid waste.

The Btu content of municipal solid waste is well-documented. The waste Btu content is very variable and will have a dramatic impact on boiler/stoker design as well as on energy recovery revenues. Btu content is dependent on the composition of the waste stream. Typically, MSW exhibits a range of 2,500 to 8,500 Btu per pounds of waste. The range is dependent on the definition of waste, and varies from 8,500 Btu/lb for trash – defined as 'highly combustible waste, paper, wood, cardboard, including up to 10 per cent treated papers, plastic or rubber scraps; commercial and industrial sources', to 2,500 Btu/lb for garbage – defined as 'animal and vegetable wastes, restaurants, hotels, markets; institutional, commercial, and club sources'. Higher moisture and inert content will have higher detrimental impact on the final MSW heating value.

Two values can be computed as higher heating values (HHV) and lower heating values (LHV). LHV does not take into consideration the energy contained in condensation of the water vapor. Since typical incineration processes do not recover the energy from vapor, the industry (except in the USA) generally uses LHV.

4.5.7 Practical methods to evaluate waste heating value

As noted in the paragraphs below, there are a number of methods used by WTE consultants in determining the LHV of MSW waste streams for a proposed facility.

Method 1 – This method uses an actual full-size WTE boiler to process large samples of MSW (i.e., 100 tons) and measures the heat recovered by computing

Table 4.4 Detailed composition results, overall waste stream, American Samoa, 2009

Material	Tons	Mean%
Paper	***5,531***	***26.4***
Newspaper	268.27	1.28
Magazines and catalogs	36.33	0.17
Cardboard and kraft	4,003.23	19.10
Boxboard	433.21	2.07
Phone books and directories	0.00	0.00
Ledger	2.74	0.01
Other office paper	42.06	0.20
Mixed-low grade paper	403.58	1.93
R/C paper	341.64	1.63
Glass	***722***	***3.4***
Clear bottles and containers	168.22	0.80
Colored bottles and containers	543.10	2.59
Other glass	11.07	0.05
Metal	***1,659***	***7.9***
Ferrous cans	1,061.31	5.06
Major appliances (white goods)	25.90	0.12
Other ferrous	82.44	0.39
Aluminum cans	209.14	1.00
Other non-ferrous	79.02	0.38
Covered electronic devices	17.37	0.08
Other e-waste	8.44	0.04
R/C metal	175.29	0.84
Plastic	***2,680***	***12.8***
HDPE – natural	189.33	0.90
HDPE – colored	174.14	0.83
PETE	256.03	1.22
Expanded polystyrene	237.92	1.14
Miscellaneous plastic containers	68.13	0.33
Bag film plastic	825.62	3.94
Film plastic	532.89	2.54
Durable plastic	76.17	0.36
R/C plastic	319.94	1.53
Organics	***4,110***	***19.6***
Food waste	791.83	3.78
Fish waste	919.36	4.39
Fish meal	0.00	0.00
Yard waste	2,389.70	11.40
Other organic waste	8.74	0.04
Carpet and padding	0.00	0.00

(*Continued*)

Table 4.4 Detailed composition results, overall waste stream, American Samoa, 2009—Cont'd

Material	Tons	Mean%
Other organic	0.00	0.00
Textiles	***875***	***4.2***
Textiles	875.49	4.18
Construction/demolition	***594***	***2.8***
Inerts	0.00	0.00
Rock, soil and fines	6.90	0.03
Wood/lumber	229.63	1.10
Treated wood	33.83	0.16
Gypsum board	190.26	0.91
R/C construction and demolition	133.18	0.64
Hazardous	***8***	***0.0***
Paint	0.00	0.00
Vehicle and equipment fluids	0.00	0.00
Waste oil	0.00	0.00
Batteries	3.89	0.02
R/C hazardous	4.14	0.02
Special	***1,416***	***6.8***
Treated medical waste	0.00	0.00
Untreated medical waste	120.28	0.57
Diapers	1,075.80	5.13
Bulky items	181.83	0.87
Tires	10.51	0.05
R/C special waste	27.87	0.13
Mixed residue	***3,364***	***16.0***
Mixed residue	3,363.91	16.05
Overall weight	***20,960***	

Source: Reference [2].

a detailed heat and mass balance to determine the LHV. Typically, this method needs to be applied for each identified 'season' throughout the year to produce the best results. It usually produces results that are about 5 per cent accurate.

Method 2 – Under this method, moisture and inert content of a few large samples are measured for each season. Then, using a modeling tool with a large database, a fairly accurate LHV is computed from these two parameters. Examples of LHV versus moisture and inert content are shown in Table 4.6. This method usually produces results that are about 10 per cent accurate.

Method 3 – The first step in this method is to measure solid waste composition (paper, food waste, plastic, etc.) for each season to explain variability. Table 4.5

Table 4.5 Examples of waste composition

	Percentage of waste				
	China (1993)		**Manila (1997)**	**22 European Countries (1990)**	
	Range	**Mean**	**Mean**	**Range**	**Mean**
Food and organic waste	40.1–71.2	46.9	45.0	7.2–51.9	32.4
Plastics	0.9–9.5	4.9	23.1	2–15	7.5
Textiles	0.9–3.0	2.1	3.5	n.a.	n.a.
Paper and cardboard	1.0–4.7	3.1	12.0	8.6–44	25.2
Leather and rubber	Neg	Neg	1.4	n.a.	n.a.
Wood	Neg	Neg	8.0	n.a.	n.a.
Metals	0.2–1.7	0.7	4.1	2–8	4.7
Glass	0.8–3.4	2.2	1.3	2.3–12	6.2
Inerts (slag, ash, soil, etc.)	14.0–59.2	40.2	0.8	Neg	Neg
Others	Neg	Neg	0.7	6.6–63.4	24.0

Neg, negligible; n.a., not applicable.
Source: Reference [3].

above shows samples of heating values for solid waste components. Using a modeling tool with a large database, the LHV can be calculated from the seasonal composition data. This method usually produces results that are about 15 per cent accurate.

Method 4 – This method consists of having a solid waste industry expert with many years of WTE experience to physically see a large sample of proposed waste stream at the landfill, witness solid waste collection by generator type (private, commercial, light industry, etc.), and evaluate the impact of seasonal variation on the waste stream. This method usually produces results that are about 20 per cent accurate.

Method 5 – This method consists of sending a small sample to a laboratory to compute a 'measured' LHV. Typically, this is the preferred method recommended by clients and consultants because of the perceived scientific element of this method. Unfortunately, experience has proven that this oftentimes is the least accurate method

Table 4.6 Worldwide variability in waste Lower Heating Value

Region	USA	Europe	China	Middle East
Inert	26	21	16	20
Moisture	31	37	57	46
LHV (kcal/kg)	2,100	2,000	1,100	1,500

because its results are hugely dependent on the content of the small sample and of the laboratory methodology. This method usually produces results that are about 50 per cent accurate.

4.5.8 Variability in waste LHV

Table 4.6 shows the variability in worldwide measurements of LHV, moisture, and inert percentages for the United States, Europe, China and the Middle East.

References

[1] Robinson William D, editor. The Solid Waste Handbook. NY: John Wiley & Sons; 1986.

[2] SCS Engineers. Waste Characterization and Waste to Energy Facility Study, 2009. American Samoa Power Authority

[3] World Bank. Municipal Solid Waste Incineration. World Bank Technical Guidance Report. Washington DC: The World Bank; 1999.

5 Waste flow control

Chapter Outline

5.1 Introduction

One of the more critical issues facing public officials pursuing WTE facilities is what is commonly termed 'waste flow control [1].' In essence, each community must be able to assure those who will be operating its WTE facility and the financial underwriters for such a project that the solid waste generated from residential, commercial, and industrial establishments within the community will be available on a long-term basis to support a WTE facility. Without strong control of the solid waste stream, there is the potential for diversion of solid waste from the community's facility. This would be an unacceptable situation because the revenues from tipping fees and the sale of electricity, steam, or recovered materials are used to finance the construction and long-term operation of such facilities.

Waste stream control has been an issue of controversy in recent years in the United States between the WTE industry and local governments on one hand, and the solid waste haulers and the waste recycling industry on the other hand [2]. This latter group has argued against the imposition of monopolistic waste flow control by local government for WTE facilities since this would interfere with interstate commerce and severely restrict their long-term financial liability by restricting their continued access to recyclable materials taken from the waste stream [3]. Representatives for the group have asserted that recycling of materials from a community's waste stream would be beneficial rather than detrimental to the financial integrity of WTE facilities because the size and capital costs of such facilities could be reduced through initiation of flow reduction programs [4].

Many communities, including the investment community, have rejected this argument; representatives for these groups have argued that the financing of WTE facilities cannot take place without the long-term assurance on the part of government

Waste-to-Energy. DOI: 10.1016/B978-1-4377-7871-7.10005-X

that a community's solid waste is committed for delivery to the WTE facility [5]. Without such assurance, the investment community has asserted that the interest rate for project financing would increase dramatically. Furthermore, some representatives of local government have asserted their rights to prohibit scavenging of materials at the curbside because of public health and safety considerations. In recent years, some communities have attempted to take a middle course by enacting waste flow ordinances with commitment for WTE facilities, while at the same time encouraging the development of a strong recycling industry in their community [6]. Waste reduction and the development of WTE projects need not be incompatible.

5.2 Flow control mechanisms

This chapter will discuss the three general waste flow control mechanisms prevalent in the United States:

1. legislative supplemented by enforcement;
2. contractual; and
3. economic or cost incentives.

5.2.1 Waste flow control through legislation/regulation

Local government in the United States can exercise some type of legal or regulatory authority over the collection, removal, and disposal of solid waste in its area of jurisdiction. Courts have long upheld the right of governments to adopt reasonable regulations in this area since all property rights were considered to be superseded by local government's police powers [7]. Most of the court cases involving solid waste were decided by jurists at the turn of the century on the premise that regulatory authority was essential to public health and safety, since without such control solid waste would become a nuisance to neighboring property owners.

When WTE projects were implemented in the United States in the early 1980s, there was some concern that solid waste flow control would be deemed unconstitutional. Perhaps the most important legal decision during that era regarding solid waste flow control by local government involved the City of Akron, Ohio. Under the terms of the bond convenants for the US$46,000,000 bond issue of its 1,000 tons per day WTE facility, the City of Akron was obliged by its bond underwriters to enact an ordinance that would:

- Guarantee a supply of solid waste to the facility by prohibiting the establishment of alternative solid waste disposal sites;
- Require all garbage collectors within Akron and Summit County, Ohio to deposit all waste acceptable for disposal at the plant (including recyclables); and
- Require all collectors to pay a tipping fee when they deposited solid waste at the facility.

Violation by a solid waste collector of this city ordinance would result in a possible loss of license and make the perpetrator subject to criminal penalties.

Prior to enactment of this ordinance, private collectors in the City of Akron and Summit County had been able to shop around for solid waste disposal sites with the best disposal price, and they could even recover and sell valuable recyclables from the solid waste stream before taking the remainder to the landfill. The imposition of waste flow control in the Akron, Ohio area interfered with the previous operations of the private collectors and landfill operators by substantially reducing their incomes.

In *Glenwillow Landfill, Inc. vs City of Akron, Ohio*, 485 F. Supp. 671 (ND Ohio 1979), these groups argued at the Federal District Court level that the City solid waste control ordinance:

- Violated due process;
- Took private property in violation of the Fifth Amendment of the US Constitution;
- Illegally restrained interstate commerce allowed under the Commerce Clause of the US Constitution; and
- Violated the Sherman Anti-Trust Act.

In ruling for the City of Akron, the court found that the ordinance was a proper exercise of the City's police powers, and as such, did not result in a taking for which compensation must be paid under the Fifth Amendment of the US Constitution.

The District Court's ruling appealed to the US Court of Appeals, Sixth Circuit. In *Hybrid Equipment Corp. vs City of Akron*, Ohio, 654 F. 2d 1187 (6th Circuit 1981), the Circuit Court upheld the City of Akron's waste control ordinance. Citing two US Supreme Court cases (*California Reduction Company vs Sanitary Reduction Works*, 199 US 306, 26 S. Ct. 100, 50 L.Ed 204, 1905, and *Gardner vs Michigan*, 199 US 325, 26 S. Ct. 106, 50 L.Ed 212, 1905), the Court found that the City had not violated the due process and taking clauses of the US Constitution, and the ordinance was a proper exercise of the traditional exercise of local governments police powers. The Court also ruled that the City's actions did not seriously burden interstate commerce, and that the City was exempt from the Sherman Anti-Trust Act. The plaintiffs in this case then appealed to the US Supreme Court.

However, before this case reached the Supreme Court, the Court had handed down a ruling in *Community Commmunications Company, Inc. vs City of Boulder, Colorado*, 455, US 40, 102 S. Ct. 835, 70 L. Ed 2d 810 (1982) holding that a municipality can be held responsible for violations of the federal antitrust laws unless it is acting pursuant to: '...a clear and affirmatively expressed State policy' permitting such restraint of trade. In light of its decision in the Boulder case, the Supreme Court overturned the judgment against the plaintiffs in the Akron case (solid waste haulers) and sent the case back to the Circuit Court to reconsider the issue of State action exemption for local government under the Sherman Anti-Trust Act. The Sixth Circuit remanded this case to the Federal District Court for disposition.

In reviewing the facts of the case, the District Court ruled that the City of Akron was exempt from antitrust liability because the City was acting in furtherance of 'clearly articulated and affirmatively expressed' policies of the State of Ohio in the financing of waste disposal facilities. The Court found that the State Legislature of Ohio had contemplated the use of anti-competitive measures to ensure the financial viability of its waste disposal facilities. Consequently, the Court reasoned that, as long as local government was acting pursuant to a clearly articulated and affirmatively expressed State policy that indicates an intent of the legislature to displace competition with regulation, local government would be exempt from antitrust liability. Furthermore, the Court, relying on *Town of Hallie vs City of Eau Claire* (No. 82-1715, Slip Op, 7th Circuit Court, February 17, 1983) held that the active State supervision requirement for antitrust immunity does not apply to municipalities engaged in a traditional municipal function authorized by the State.

Prior to the final disposition of the Akron case, the financial and legal advisors to communities who were hoping to implement WTE facilities insisted that some sort of legal mechanism be used to confer State immunity from antitrust actions upon local government. To achieve such immunity, however, it was believed that local government needed to meet the 'active' State supervision test. Some local governments have attempted to demonstrate such State oversight of their WTE programs by having the State, through special legislation, officially delegate the power of supervision to them. Others, through special legislation, have reaffirmed the existing State supervisory powers over their WTE systems, including the issuance of permits and periodic reviews.

The Akron decision makes it clear that waste flow control ordinances for WTE facilities must be carefully drafted by local government. They must balance the needs of government to assure secure waste supplies for its WTE facility against the legitimate economic concerns of collectors and the recycling industry to remove recyclable materials from the waste stream. WTE facilities and recycling need not be incompatible. Many local governments, while enacting strong flow control ordinances, now permit the recovery of recyclable materials in the waste stream. However, solid waste flow control continued to be the subject of litigation in the United States for another decade.

In 1994, the US Supreme Court issued a far-reaching and landmark legal opinion regarding solid waste flow control. In *C&A Carbone, Inc., et al., vs Town of Clarkstown, New York* (1994) the Court deemed a local flow control ordinance in New York unconstitutional because it violated the Commerce Clause of the US Constitution by driving competitors from outside the local market (including out-of-state competitors). The Town of Clarkstown had hired a private contractor to build a waste transfer facility and enacted an ordinance requiring that all solid waste generated within the Town be directed to the transfer facility (with tipping fees higher than the disposal costs of the private market). The Town had financed the transfer station and planned to be paid back from tipping fees generated there. The Supreme Court struck down the ordinance, stating that solid waste was a commodity in commerce and that the Commerce Clause invalidates laws that discriminate against such commerce because of its origin or its destination out-of-state. The Court found that flow control laws 'deprive competitors, including out-of-state firms of access to a local market.' (*C&A Carbone, Inc.*, at 386.)

The Carbone decision was applied to solid waste management in New Jersey in 1997 in the case of *Atlantic Coast Demolition & Recycling, Inc. vs. Board of Chosen Freeholders of Atlantic County, et al.*, 112 F.3d 652 3d Cir. 1997 ('Atlantic Coast'). New Jersey is one of the most densely populated States in the nation. Coupled with the scarcity of available land for landfill facilities, its large waste-generating population, and the State's geographic location relative to other large metropolitan centers (Philadelphia and New York), enormous resources have been expended over the last three decades to ensure safe and effective solid waste disposal capacity for refuse generated by its citizens. As such, 13 new solid waste WTE facilities needed to be constructed; while at the same time, the State and local communities had to grapple with the closure of more than 500 landfills with poor environmental records that failed to meet increasingly stringent regulations that had been imposed nationwide by the US Environmental Protection Agency (EPA). At the same time, the State Legislature was addressing New Jersey's long history of anti-competitive conduct in the solid waste industry through rate regulation and State certification of waste collectors.

During the late 1970s, the State adopted what has been termed the 'State self-sufficiency policy' in which the New Jersey Department of Environmental Protection (NJDEP) refused to approve solid waste district plans from each county, which proposed long-term out-of-state disposal arrangements because it was believed by NJDEP that these arrangements were not reliable over the long-term. Instead, the NJDEP required in-district or in-state disposal as long-term solutions. As a consequence, counties in New Jersey financed the construction of their solid waste facilities, pursuant to their solid waste plans by issuing revenue bonds (almost US$1.2 billion collectively by the 1990s), which were assured by the guaranteed flow of waste to their publicly-owned facilities. Unfortunately, the legal uncertainty surrounding the State's 'self sufficiency' policy and permissible governmental regulation of solid waste collection and disposal has required changes to this State policy.

Atlantic Coast was a Pennsylvania waste hauler and transfer station operator. It accepted construction and demolition debris at its transfer station in Philadelphia, separated out the recyclable materials, which amounted to less than 20 per cent of the total by weight, and shipped the residue to various landfills for disposal. It sought to obtain access to construction and demolition debris generated in New Jersey, but it was unsuccessful in having its transfer station included as an authorized facility in any New Jersey district waste management plan. Since New Jersey's waste management regulations required shipping all non-recyclable solid waste collected in New Jersey to the transfer stations designated for the districts from which the waste had been taken, Atlantic Coast's only course consistent with New Jersey law would have been to return non-recyclable waste to the designated facilities for processing or to pay a compensating fee. Rejecting those expedients as too costly, Atlantic Coast commenced an action in the United States District Court for the District of New Jersey, challenging the constitutionality of New Jersey's solid waste flow control regulations.

In this landmark case, the Third Circuit of the US Court of Appeals reversed the finding of the District Court and found that New Jersey's system of regulatory flow violated the interstate commerce clause. The district solid waste management plans that mandated all solid waste to be delivered to designated facilities in districts throughout New Jersey were determined to discriminate against out-of-state competition. In striking down New Jersey's restrictions barring competition from out-of-state facilities and operators, the Third Circuit limited its finding of unconstitutionality to the State's Solid Waste Management Act., the regulation that codified waste flows contained in the district plans.

As a consequence of the *Atlantic Coast* decision, each district in New Jersey has struggled to address this new legal landscape regarding solid waste flow control. Tipping fees in the State fell literally overnight due to increased competition from out-of-state haulers. Those districts that contracted with private waste haulers modified their systems through re-bidding their waste contracts open to both in-state and out-of-state bidders. Some district-sponsored or -supported facilities began charging a market or below-market tipping fee to attract waste and subsidized the operating costs and debt service through use other public funds (Environmental Investment Charge). Lawsuits were filed challenging the implementation of such charges in several districts.

Those communities that had expended public funds to construct solid waste facilities that relied on waste flow control faced a financial dilemma. This 'stranded debt issue' ultimately required the State to step in and subsidize the debt payments of certain counties and write off certain solid waste-related State loans in order to prevent default and the difficulties that could result for public agencies statewide that seek to raise capital. In 1998, New Jersey voters approved Ballot Question No. 3, which authorized the writing-off of US$103 million in loans to seven counties, and set aside US$50 million in bond money to help solid waste facilities avoid bankruptcy.

In April 2007, the US Supreme Court made a landmark decision regarding solid waste flow control in *United Haulers Association vs Oneida–Herkimer Solid Waste Management Authority*, 550 US (2007). United Haulers sued the New York counties of Oneida and Herkimer, claiming that county ordinances regulating the collection, processing, transfer, and disposal of solid waste violated the Commerce Clause. The flow control regulations enacted by the two counties required all solid wastes and recyclables generated within Oneida and Herkimer counties to be delivered to one of several waste processing facilities owned by the Oneida–Herkimer Solid Waste Management Authority. United Haulers had argued that these ordinances burdened interstate commerce by requiring garbage delivery to an in-state facility, as this restriction necessarily prevented the use of facilities outside the Counties and diminished the interstate trade in waste and waste disposal services. United Haulers had submitted evidence that the flow control ordinances increased the cost of waste transport disposal from between US$37 and US$55 per ton without flow control to US$86 per ton with flow control.

By a 6 to 3 decision, the Court ruled in favor of the two counties upholding solid waste flow ordinances that required waste haulers to deliver their trash to a publicly operated processing site. The justices disagreed with United Haulers, stating that the counties' flow control ordinances, 'which treat in-state private business interests exactly the same as out-of-state ones, do not discriminate against interstate commerce.' In stark contrast to the previous *C&A Carbone, Ltd.*, decision, a majority of the justices argued that this case was different because it dealt with a publicly owned solid waste facility, which benefits the local 'government's important responsibilities including protecting the health, safety, and welfare of its citizens.' The precise scope and impact of the Supreme Court's decision is currently unclear.

Since the decision was announced, a few counties and municipalities have begun adopting solid waste flow ordinances. If this trend continues, this may limit the potential availability of solid waste in the market area surrounding the Authority. Notwithstanding, the scope of the public sector exceptions in the United Hauler case and their application to specific factual circumstances involving solid waste management is expected to be the subject of further litigation in the federal courts.

In recent years, a number of rail-based transfer stations have been sited in New Jersey to receive and transport municipal solid waste. The developers received approval from the Surface Transportation Board (STB). Congress created the STB in 1995, in an effort to generate uniformity and consistency in the regulation of rail transportation. The STB has exclusive jurisdiction over railroad operations, and, with that, the power. It has proposed exempting rail-based solid waste transfer activities from State and local permits, and ruled that certain waste transloading activities that take place on or near railroad rights-of-way constitute 'transportation by rail carrier,' and are therefore exempt from State laws governing solid waste management.

The NJDEP and other solid waste professional organizations have argued against the STB preemption as applied to trackside solid waste facilities, because it removes critical controls that ensure that these operations are conducted in a manner that will protect the environment and public health and safety in all communities where they are located. For these reasons, these groups support measures to end the STB's authority to exempt railroad-related solid waste facilities.

In November 2008, The President of the United States signed legislation effectively ending the federal exemption of all rail facilities that handle solid waste. The Clean Railroads Act of 2008 (Public Law 110-432) permanently closed the federal loophole and allowed States to regulate solid waste facilities on rail property for environmental, health, and safety reasons. This legislation was incorporated into the comprehensive railroad safety bill.

5.2.2 Contractual control of waste stream

Rather than resorting to the enactment of waste stream control legislation, local government can assure adequate quantities of solid waste for its WTE facility through contractual controls. This is accomplished when local government enters into long-term contracts with other local governments and private collectors to deliver solid waste to a WTE facility. This method of voluntary contractual commitments can be particularly effective to secure an adequate core of solid waste for the facility.

This technique has been successfully utilized in several WTE facilities in recent years to assure long-term waste supplies. For example, refuse for the Northeast Massachusetts Resource Recovery Project located in North Andover, Massachusetts is delivered to the facility by 22 municipalities that have signed 20-year 'put-or-pay' agreements for waste disposal and eight commercial haulers who have signed private hauler agreements. These 'put-or-pay' agreements require each community to deliver a guaranteed annual tonnage, which can be adjusted yearly within certain limitations. These communities are assessed penalties for shortfalls or excesses below or above their contractual guarantees. Private haulers are also assessed penalties if their deliveries are below or above their contractual guarantee. In this way, the WTE facility can capture adequate quantities of solid waste to meet its financial commitments to the local communities, the facility operator, and the bondholders.

5.2.3 Economic incentives for waste stream control

Waste stream control can also be achieved by local government through economic incentives. For example, the operator of a WTE facility can attract solid waste from both public and private collectors by charging a zero or lower tipping fee than for alternate disposal methods such as sanitary landfills. In this case, private haulers would be attracted to the facility since they would have no economic incentive to dispose of their solid waste at less convenient sanitary landfills elsewhere.

In order to accomplish this type of economic control over solid waste for its WTE facility, the community must be willing to subsidize the loss of project revenues with funds from some other source, such as from the general fund, a user fee, or a tax. For example, a user fee for solid waste disposal can be established for different residential, commercial, and industrial accounts whereby the proceeds from this fee could be used to offset the zero or artificially low tipping fee at the WTE facility. Some communities have also used revenue from property or other local government taxes to subsidize the tipping fee at their WTE facilities. Use of property taxes, however, may be viewed by the investment community as a general obligation of the community and could result in a lowering of its bond rating.

References

[1] William F Cosulich Associates. The Integration of Energy and Material Recovery in the Essex County Solid Waste Management Program. Belleview, New Jersey: Essex County Division of Solid Waste Management; 1983.

[2] Snyder David L. Antitrust Law and Solid Waste Management: The Municipal Perspective. In: Proceedings of the Governmental Refuse Collection and Disposal Association, August 16, 1982 in Dallas, Texas. Washington: Governmental Refuse Collection and Disposal Association; 1982.

[3] Felago Richard T. Waste Stream Assurance: Key to Resource Recovery. Solid Wastes Management; July 1982.

[4] Franklin William E, Franklin Majorie A, Hunt Robert. Waste Paper: The Future of a Resource 1980–2000. Prairie Village, Kansas: Franklin Associates, Ltd; 1982.

[5] Personal communication from Charles Citrin, Sparber, Shevin, Rosen, Shapo and Heilbronner on September 17, 1982.

[6] Kovacs William L. Flow Control: An Unnecessary Constitutional Conflict in Managing Solid Waste. Environmental Analyst 1982. June 3–7.

[7] Spiegel David R. Local Governments and the Terror of Antitrust. American Bar Association Journal 1983;69:163–6.

6 Selecting the facility site

Chapter Outline

6.1 Introduction

The siting of a major public facility, especially a WTE project, is not a simple task, particularly when such facilities are often located in highly developed and environmentally conscious communities. Many technical, environmental, and social (institutional) issues must be considered. The site selection process is complex, not only requiring the project developer to identify a site that minimizes adverse environmental impacts and can accommodate the operation of a WTE facility, but also requiring the development of specific site evaluation criteria that have a reasonable chance of public acceptance [1,2]. To achieve this latter objective, the site evaluation criteria so devised must be well-documented and carried out in a uniform and consistent manner.

This chapter describes a generic site selection process that can serve as a model for siting efforts of WTE projects and avoid some of the major pitfalls. The method described can allow project developers to identify feasible sites; to eliminate the less suitable ones; and to recommend the best site(s) in a detailed and objective way [3]. Furthermore, these siting methods can help enable communities to win public support for such sites, which is the key to the successful implementation of any WTE project.

Waste-to-Energy. DOI: 10.1016/B978-1-4377-7871-7.10006-1

6.2 The site selection process

The site selection process for a WTE project requires several stages of analysis (Figure 6.1). These steps are in essence screening stages within the selection process as they progressively narrow the criteria for analysis and evaluate more detailed data.

6.2.1 Evaluation criteria

The first step in the site selection process is to identify and document the criteria [5]. It is necessary that all of the criteria important to the siting of a facility receive balanced consideration. In order to maintain clarity in this effort and to provide a uniform method of reviewing and screening sites, the following three broad categories of site evaluation criteria have proven to be extremely useful for many projects:

- Technical considerations;
- Environmental considerations; and
- Social (institutional) considerations.

By structuring the evaluation this way, project developers can consistently analyze proposed facility sites and ensure that critical issues are not overlooked.

Each major division can then be subdivided into smaller evaluation criteria for more specific, detailed appraisals. The criteria under each broad category vary from one area to another based on local situations. Some typical criteria are:

- Technical considerations:
  - Site drainage
  - Foundation suitability
  - Size and shape of site

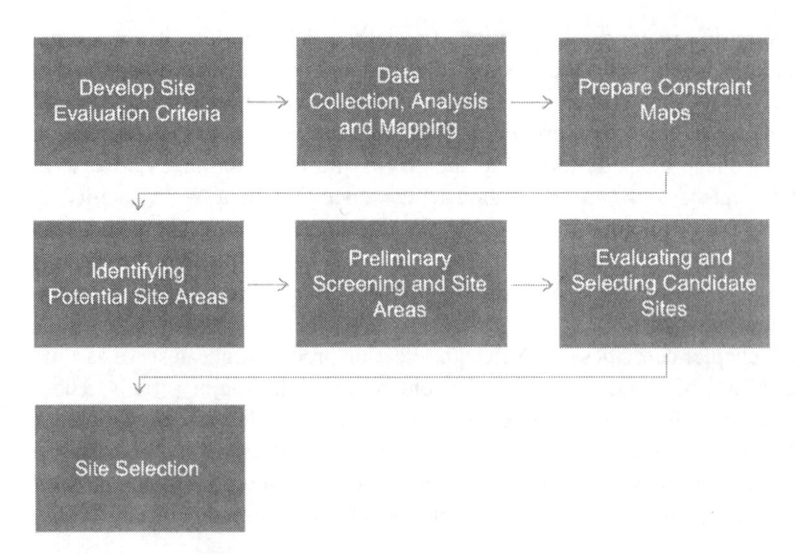

Figure 6.1 Flow chart of the site selection process.

- Accessibility (e.g., highway, railroad, barge access.)
  - Location
  - Utilities;
- Environmental considerations:
  - Air quality
  - Water quality
  - Biological resources (e.g., fauna and flora);
- Social (Institutional) considerations:
  - Surrounding land uses
  - Permitting considerations
  - Land ownership
  - Cultural resources.

By applying these criteria, candidate sites can be identified and a preferred site selected. The specific features of some of the major evaluation criterion are briefly described in the following sections.

6.2.1.1 Technical considerations

Site drainage. Site drainage is an important design consideration for a WTE facility. While buildings, roadways, and ancillary facilities can be shielded by natural or artificial features to assure protection against flooding, surface runoff must be directed into nearby watercourses to be carried offsite or retained onsite. Small watercourses can often be rerouted and stormwater detention/storage basins can be developed based upon the design requirements of the local government entity. Constructing these improvements usually results in increased development costs for a site. In addition, flood control ordinances in some communities may prohibit construction of facilities in floodplains. Locating facilities in such areas may also increase the cost of insuring the WTE project, if flood insurance is available.

Foundation suitability. WTE facilities generally require large and complex buildings, because equipment such as boilers, generators, and air-pollution control devices are unusually heavy. This necessitates that the site for such facilities has stable soils for construction of foundations. While unstable geological conditions can be overcome through the design and construction of more complex foundations, such conditions could preclude the use of an otherwise attractive site because of the additional expenses involved.

Similarly, high groundwater conditions or shallowness of the site to bedrock could result in more extensive and complex foundation designs. For example, large WTE facilities are often designed with a storage pit of sufficient capacity to store several days of solid waste fuel. Such storage areas are usually excavated below existing grade to provide the necessary storage volumes. Thus, sites with high groundwater conditions may require that either the entire structure be raised through the use of fill as an alternative, or a shallow tipping floor be utilized. In either case, such designs may significantly increase the construction and operation of the facility as compared to sites not requiring such innovative foundation designs.

Size and shape of site. The size and shape of a site required for a WTE facility is project-specific and could vary from as small as five acres to as large as 50 acres. For example, the size of a site can be dependent on the following key factors: the type of technology used to process the solid waste (e.g., mass burn, RDF); the quantity of refuse to be processed; and the location of and access to the site in relation to neighboring land uses. Consequently, land requirements for WTE facilities can vary greatly from one community to another. For example, a site surrounded by heavy industrial land uses may require minimal buffering, thus reducing the acreage needed for the project. However, sites bordering light industrial and residential land uses, or those having flat topography, may require significantly more acreage to provide buffer zones between the plant and its neighbors. In addition, local traffic and road conditions may also have an impact on the overall size of the site because special access road configurations may be required to adequately handle the number of vehicles entering and leaving the facility. This could add additional acres to a community's preferred parcel size.

Accessibility. An operating WTE facility generates significant numbers of vehicles, which deliver solid waste (to be processed and hauled away), recovered materials, and residues. Due to the size and numbers of these vehicles, and to minimize potential traffic congestion and accidents, it is preferable that access to a facility site be from major highway and rail systems and not through high-density residential areas. The ability of the local road network nearest a proposed site to handle the flow of solid waste vehicles safely and efficiently must be examined. Project developers should evaluate such factors as: road widths; structural capabilities of roadways and bridges; weight limits; height restrictions; speed limits; and grades. The purpose is to determine whether existing roadways can safely carry an increased vehicle traffic load. The costs of providing these services can be useful information to assist in the ranking of candidate sites.

Location. The location of a site for a facility is an important evaluation factor. Such facilities should be located within reasonable distances from the solid waste collection area; the energy sales market; needed utilities such as electricity; cooling and boiler makeup water; and the sanitary landfill that will accept the ash residues or bypass waste.

The ideal position for a WTE facility is a site located close to the center of the waste generation centroid, adjacent to a sanitary landfill and energy customer, and within the local government's utility service area. It is unlikely that sites meeting all such requirements for WTE projects can actually be found. Consequently, project developers must often evaluate the degree to which these requirements are met by the sites under analysis.

Foremost among the requirements for a WTE facility, is its proximity to a proposed electricity or steam customer. The costs of constructing electrical transmission lines are often prohibitive for most WTE projects. Thus, the most economical locations for a facility are sites near large electrical transmission lines or existing electric utility substations. In the case of facilities proposing to export steam to an industrial or institutional customer, sites located near an existing steam loop or the customer itself are preferable to minimize the costs of expensive steam distribution lines and their associated equipment.

Another major requirement for a WTE facility is its location to the existing and future solid waste collection area of the municipality. The closer the facility is to the solid waste generation center, the lower the cost will be to transport the solid waste to the facility. Similarly, sites located in a line between the center of solid waste generation and the sanitary landfill, which will accept the plant's ash residue, have a higher comparative advantage since haul costs of the ash residue can be minimized. However, WTE processes greatly reduce the volume of the community's solid waste by a factor of nearly 90 per cent, so it is generally more economical to locate the facility near the center of the community's solid waste generation rather than closer to the ash residue landfill [6]. Unfortunately, for many communities, the center of solid waste generation is usually found in their most developed and populated areas, making siting in these areas difficult.

Utilities. WTE facilities can consume substantial quantities of electric power, water, and natural gas, and require telephone, wastewater, and emergency services such as fire and police protection and ambulance services. Sites located where such utility services are already available are preferable to those where these services must be provided at considerable expense to the project. Some utilities such as a telephone service are readily available in most communities and can be easily extended, while other utilities such as water and wastewater services may be unavailable in some communities.

Electricity, water, and wastewater services are generally the major utility service needs of WTE facilities. Such facilities can consume substantial amounts of electric power principally at times of plant startup and outages. Electric service can usually be provided to WTE plants at reasonable expense in most communities. An electric transmission line or a substation must be located nearby to deliver the facility's energy output.

A sanitary discharge system is a requirement for the proper operation of a WTE facility. Liquid wastes usually result from many plant operations such as: boiler blowdown; water pretreatment for the boiler; and the normal sanitary requirements of the plant work force. The community's sewer system is the preferable discharge system for most sites. However, there may be instances where such services may be unavailable to handle the flow of the plant because of prior commitments. This may mean that project developers would have to consider on-site treatment and discharge, thereby increasing the potential cost of the facility.

A source of water for cooling and process needs is another major utility required for a WTE plant. Potable water from a public system or an on-site well is usually necessary for normal sanitary needs. Non-potable or recovered water from wastewater treatment plants or nearby rivers can be used for other facility operations such as evaporative cooling, boiler feedwater makeup, and fire protection. The cost of providing this water service to the plant results from either extending existing water lines, or drilling new on-site wells. This can have an impact on the desirability of one site against another that requires minimal expenditure for water services.

6.2.1.2 Environmental considerations

Air quality. The impact of a WTE plant at a particular site on the local or regional air quality is an important consideration in site selection.

WTE facilities incorporate some form of combustion process that results in various gaseous and solid emissions to the atmosphere. Good combustion control and the addition of air pollution control equipment, such as electrostatic precipitators, bag houses, and acid gas scrubbers, will help minimize the overall air pollution potential of a proposed facility, although there will still be some quantity of air emissions which could degrade the existing local and regional air quality. Areas designated by regulatory officials as not meeting existing standards for specified air pollutants will generally require more expensive air-pollution control equipment than plants located on sites in areas designated as attaining these regulatory standards.

Computer modeling of the area's air quality and meteorological data can go a long way to help evaluate the potential impact of the facility upon ambient air quality. Such models assist in determining whether the plant can meet minimum regulatory air emission standards with respect to a particular site's specific configuration. In addition, knowledge of an area's atmospheric flow characteristics can help predict whether normal wind patterns will assist in dissipating the plant's emissions away from sensitive human receptors.

Water quality. Water quality is also a very important issue in the siting of WTE facilities. WTE facilities utilize significant quantities of water for cooling and process needs. Project developers should consider the impact of the eventual disposal of liquid wastes from a WTE facility upon the water quality of nearby bodies of water and groundwater aquifers. Some states have recently considered promulgating stringent regulations restricting the development of certain land areas, located near designated high-quality waters, for construction of certain public works projects such as wastewater treatment plants and solid waste facilities.

Biological resources. There are a number of unique flora and fauna species protected by federal, state, and local regulations. It is important during the initial screening of sites for a WTE facility to identify the habitats of these threatened or endangered species to ensure that these areas be avoided for development.

6.2.1.3 Social considerations

Surrounding land uses. The compatibility of a WTE facility with its surrounding land uses is an important consideration in siting. An operating WTE facility can exhibit all the indications of an industrial-type of activity. It can generate significant volumes of truck traffic; has the potential to emit noise, odors, and dust; and may, because of its building height and configuration, suggest the visual appearance of a heavy industry. This is not to suggest that these potential impacts could not be overcome with appropriate mitigative measures. For example, the visual appearance of such facilities can be made compatible with many land uses through the judicious use of landscaping, buffer zones, and architectural materials such as glass, brick, and colored metal panels.

Early determination of land use incompatibility can eliminate significant project delays at later phases of project implementation. For example, areas near airports require special attention since the Federal Aviation Administration (FAA) regulations limit construction (particularly height) in or near airport runway approaches. Distance from airport runways and elevation of ground surface should be documented to determine whether a stack might interfere with aircraft navigation. Building height restrictions are in place in many communities around all the airport facilities, often through strict zoning regulations.

Permitting considerations. The number of permits and the length of time needed to acquire them for a WTE facility can be an important factor in the successful implementation of a project. Permitting delays due to the complexity of the regulatory process can impact a project's financial success. Although some permits will generally be required regardless of site location, there are other permits that are applicable based upon specific site conditions. At the outset of a project, it is critical that both the potential number of permits and estimated length of time required to obtain those permits are evaluated for the specific candidate sites. Under this criterion, sites that potentially require the least number of permits would be preferred as compared to those sites requiring a greater number of permits. Consequently, sites that do not contain environmentally sensitive lands that are protected under current regulations would have greater likelihood of permitting success.

Land ownership. Land ownership is an important factor in determining the availability and ease of obtaining a site for a WTE facility. Land parcels under the control of governments are preferable over privately owned lands because there is less likelihood of acquisition delays. Many governments, however, do not have parcels dedicated for use in WTE projects. In this case, privately owned lands, which have only one owner, should be preferred over sites having multiple owners due to the increased ease and speed of land acquisition.

Cultural resources. Cultural resources include such items as: archaeological areas; historic sites; and scenic landmarks. The construction of a WTE facility on or near sites having cultural significance can have both direct and indirect effects. Direct impacts can occur because of the actual construction and operation of a facility. Cultural resources can also be indirectly impacted if the presence of the WTE facility affects their use.

6.3 Site screening process

Once the evaluation criteria have been developed, and qualitative or quantitative ratings assigned to each, project developers can utilize them to screen potential sites. While there are many ways by which this site selection process can be undertaken, ideally the process includes several stages of analysis. That is, there is a progressive narrowing of criteria. The overall site selection process discussed below has worked well in the siting of WTE facilities and can be adapted to a specific project.

6.3.1 Stage 1 – Data collection and analysis

In order to apply the siting criteria in the screening process, data need to be collected and compiled to relate these abstract facts to specific places. If, for example, flooding is a key concern to a community, then data, which can detail the location and nature of flood-prone areas in the area under study, need to be collected. Many federal, state, local, and private agencies usually collect such data as part of their normal activities. Meetings should be held with these entities to ensure that the data being supplied are adequate for the community's needs.

6.3.2 Stage 2 – Preparation of constraint maps

Once the data required for the site screening process are collected, they should be illustrated in a series of maps that can help assist in further data analysis [7]. Many WTE siting studies have found it useful to portray the data in a series of constraint maps that relate similar criteria together on a single map format. For example, maps containing data on site criteria that may severely restrict development of WTE projects, can be overlayed to produce a single map which would show those areas with the most restrictive conditions for development.

Some of the criteria so portrayed may have such a significant impact upon development that the location of a WTE facility would be unsuitable, and these areas would not merit any further consideration. Other criteria may adversely have an impact on the development of a site for a WTE facility unless costly mitigative measures are incorporated in the project design. Some examples of these exclusion site criteria may be the following:

- Air quality non-attainment;
- Wetlands;
- Airport restricted zones;
- Flooding susceptibility;
- Historic sites;
- Environmentally sensitive areas;
- Prime or unique agricultural areas;
- State or local recreational areas;
- Developed lands; and
- Unstable geological conditions (e.g., sinkholes, earthquake zones).

Similarly, maps showing areas reflecting factors favoring the location of a WTE plant could also be prepared. Criteria which may favor the location of a WTE facility in an area could include the following:

- Solid waste generation centroids;
- Location near major highways and interchanges;
- Energy market areas;
- Areas near wastewater treatment plants;
- Water service areas; and
- Areas with industrial, light-industrial or commercial land uses.

6.3.3 Stage 3 – Identifying potential site areas

The constraint maps prepared in Stage 2 can be used by study participants both to eliminate areas for further consideration and to suggest general areas within a community that favor the development of a WTE facility. Potential site areas can then be identified for more detailed on-site analysis.

6.3.4 Stage 4 – Preliminary screening of site areas

The next major stage in site screening is to evaluate the potential site areas previously identified. This usually requires an on-site investigation to determine the actual site conditions at the time of the study. This on-site analysis can often eliminate some sites from further consideration because their actual conditions at the time of study may be different from those indicated on the maps that may have been completed at an earlier date. On-site analysis is often critical in rapidly developing areas, since sites often shown on maps as vacant land may already have been developed. Such field investigation can also aid in identifying more specific sites for further analysis within the large, undefined areas shown on the maps. It is important that these studies are well documented through the use of a survey form to assure opposition groups that the study was undertaken in an objective manner.

6.3.5 Stage 5 – Evaluating and selecting candidate sites

Based upon the results obtained in Stage 4, a list of potential sites can be constructed from which candidate sites may be evaluated. At this point in the site selection process, it is useful to obtain additional data on the sites, such as the number of separate parcels of land for each site, their ownership, and land prices. The sites should be revisited to verify the initial field investigation conducted in Stage 4 and to obtain any needed additional site data.

It is at this time that the sites under consideration can be rated using the siting criteria that had been developed for the study. To illustrate this procedure, each site could either be assigned a quantitative or qualitative rating for the technical, environmental, and social siting criteria. As shown in Table 6.1, for example, each site could be assigned a qualitative rating of 'good,' 'average,' or 'below average,' or some similar rating scheme. Thus, assuming that the three criteria categories have equal importance, the overall rating for a particular site would be a composite of its individual ratings for technical, environmental, and social criteria.

To reflect the relative value of each siting criteria more precisely, a weighting system can be developed. Each individual criterion or criteria category (e.g., technical, environmental, and social) can be assigned a particular weighting to indicate its relative importance. For example, technical considerations could be assigned a weighting of three; environmental considerations a weighting of two; and social considerations a weighting of one. Such weightings can be developed through discussions with the community's decision-makers, civic and environmental interest groups, and the public. By utilizing this weighting procedure, the siting team can narrow its search to relatively few candidate sites.

The first task of the team in evaluating such sites is a more detailed review of the data collected earlier, and the collection and analysis of additional data. The detailed evaluation of these remaining sites could include the following steps:

- Meetings and correspondence with property owners and other landowners adjacent to the candidate sites;

Table 6.1 Example of weighted ratings for WTE sites

Sites	Ratings			
	Technical	Environmental	Social	Composite
A	A	G	G	G, A
B	G	A	A	G, A
C	G	G	G	G
D	G	A	A	G, A
E	G	G	G	G
F	G	A	A	G, A
G	G	A	G	G
H	A	G	A	A
I	Ba	Ba	Ba	Ba
J	A	G	G	G, A
K	G	Ba	A	A
L	Ba	G	A	A
M	A	A	G	A

G, Good; A, Average; Ba, Below average.
Technical considerations given a weighting of three; environmental considerations a weighting of two; and social considerations a weighting of one.
Sources: References [1] and [4].

- Meetings and correspondence with a number of federal, state, local and private agencies to gather and evaluate data;
- Detailed site visits, including discussions with nearby residential civic associations, and officials of public institutions such as schools, hospitals, and emergency services that might be required by the WTE facility;
- Completing a preliminary air quality impact analysis at each site, including modeling of emissions and detailed permitting requirements;
- Undertaking a preliminary traffic impact assessment at each site, including modeling potential vehicle flows at different roadway segments at each site;
- Analysis of noise impacts, including measurements of existing ambient noise levels and assessments of predicted noise impacts at each site;
- Assessment of required mitigative measures to address any potential aesthetic and visual impacts at each site;
- A geological investigation at each site, including several subsurface borings; and
- Comparative cost analysis of each site, including relative differences in transporting, constructing, and facility operating costs.

6.3.6 Stage 6 – Site selection

The site evaluation criteria from Stages 1 through 5 form the basis for selection of the candidate sites. In many cases, there are few sharp differences in the overall ratings among the remaining sites themselves. Consequently, in order to make more detailed distinctions among these sites, many studies have found it necessary to regroup their evaluation categories to address key project issues such as: compatability with adjacent land uses; environmental impact; and comparative project costs. These

Table 6.2 Example of the comparative ranking of four candidate WTE sites

Selection parameters	Site F	Site H	Site B	Site P
Compatibility with adjacent land uses	2	3	1	4
Environmental impact	2	3	4	1
Comparative costs	1	3	4	2

Sources: References [1] and [4].

factors can then be assigned numerical values for each site. As illustrated in Table 6.2, a '1' could signify, for example, the most valuable ranking, and a '4' could signify the least valuable ranking. These factors could then be added together and the site with the overall lowest numerical value could then be recommended for acquisition.

Land use compatibility. Evaluation of land-use compatibility can play an important role in the final selection of a project site. This evaluation issue could include those siting criteria that may have an impact upon a proposed WTE facility with its surrounding land uses. A technical criterion, such as accessibility, could be used to determine the impact of the plant's traffic flows upon neighboring land uses. Social criteria, such as surrounding land uses, land ownership, and cultural resources, could be utilized to assess the effect of the plant on existing or proposed land uses. Permitting considerations are another criterion that would reflect the need for zoning and comprehensive plan modifications to assure the site's compatibility with adjacent or surrounding land uses. An environmental criterion, such as air quality, could be used to reflect concerns with possible facility impacts upon surrounding areas.

Environmental impact. Major environmental constraints, which would prohibit the development of a WTE facility, will have been considered in earlier stages of the siting analysis. The environmental impact review of the candidate sites in this stage, however, can focus on their relative differences. For example, the following environmental categories can be examined in detail: air quality; water quality; site drainage; and biological resources. The use of such categories can allow the evaluation to focus on the most critical environmental issues associated with a WTE facility, and on environmental features that may differ from site to site. For example, the results of the air quality impact analysis can provide additional input into the analysis by helping predict the need for additional air pollution control devices or an increase in stack height due to potential downdrafts of air pollutants at a particular site. Technical criteria, such as site drainage and accessibility, could be utilized to reflect potential impacts of flooding and traffic flow, respectively, upon the local environment.

Comparative costs. All candidate sites can be compared for costs. Thus, differences between the sites could be considered, assuming that identical facilities could be constructed at each site. This comparison could include technical criteria such as: site drainage; geological conditions; utilities; and accessibility. These would reflect relative construction costs of the plant at each site. The relative differences in the costs of hauling solid waste to each site from the center of the solid waste collection area could be assessed using the criterion of location. Environmental criteria, such as air quality and water quality, could also be considered. Based on air quality modeling, the relative cost of installing the necessary air pollution control equipment required for a facility, could be calculated for each site. Permitting considerations can also be applied under this category, both in terms of the estimated costs of obtaining permits, and in delays resulting from the acquisition of the necessary permits.

6.4 Use of Geographic Information Systems (GIS) technology in siting

GIS have been in use since the mid-1980s and, in recent years, are finding applications that are more useful in the solid waste industry. Simply put, GIS is a computerized system that integrates, analyzes, and models data from maps, surveys, photos, reports, and other sources and produces graphical maps, reports, and plans for the decision-making process. In essence, it is a visualization tool.

Assume that the community is in search of the 'best' site for the construction of a WTE facility. Before GIS, you would have to manually review hundreds of pages of parcel ownership maps in the Property Appraiser's Office; and review the maps and reports individually to identify important natural features (as described in this chapter) to avoid when choosing an appropriate site. Then, after painstakingly drawing a new map that contains all of the important siting factor data, and manually measuring the distances from all of the sensitive features, you need to be able to interpret them after you have combined everything into one drawing.

Prior to GIS technology, you would be lucky if you could achieve the siting analysis work in six months or less and often with a very high price tag. Additionally, the accuracy may be questionable because there was so much data to digest. If you had to investigate four other potential parcels for the WTE site, you could easily add another four months of work.

In contrast, if you could utilize a GIS software application, then that same assignment could be completed in about half the time as with the formerly manual method and would be likely to produce far more accurate and useful information for the decision-makers. For example, if someone wanted to know who owned certain parcels and how large they were, you could provide the answer with just a click of your mouse (Figure 6.2). Further, if you wanted to see all of the power corridors, every permitted potable well, every housing subdivision, floodplain and wetland, it would be contained in your GIS database. In essence, GIS allows you to focus on layers of information and enables you to manipulate it electronically.

Worldwide, GIS has enabled communities to visualize the data for a WTE siting process including the following aspects:

- Groundwater quality variations;
- Air emission patterns;
- Fugitive odor emissions detection monitoring;
- Collection trucks routing efficiency; and
- Three-dimensional (3-D) visualizations.

Figure 6.2 The GIS software application illustrating an aerial view.

References

[1] Camp Dresser & McKee Inc. Evaluation and Selection of a Resource Recovery Site for Hillsborough County. Florida, Tampa: Hillsborough County, Florida; 1982.

[2] Hauser Robert, Rogoff Marc J, Stoller Paul J. Site selection of a resource recovery facility in Hillsborough County, Florida. August 23–30. In: Proceedings of the Governmental Refuse Collection and Disposal Association 23rd Annual International Seminar. Silver Spring, MD: Governmental Refuse Collection and Disposal Association; 1985.

[3] Rogers John W. Resource recovery facility siting. In: Proceedings of Fourth Annual Resource Recovery Conference, March 28–29 in Washington, DC. Washington: National Resource Recovery Association; 1985.

[4] Rogoff Marc J, Clark Bruce, Gamboa Jose. Advances in technology alter how we now plan and operate solid waste management facilities. APWA Reporter 2005. March.

[5] Rogoff Marc J, Stoller Paul J, Hauser Robert. How to site a resource recovery facility. World Wastes 1984. July.

[6] Rogoff Marc J, Smith Warren N, Berry Patricia. Siting of a resource recovery facility: Hillsborough County, Florida. In: Proceedings of the Fourth Annual Resource Recovery Conference, March 28–29 in Washington, DC. Washington: National Resource Recovery Association; 1985.

[7] Wade David. Siting and the thermal energy market. In: Proceedings of Third Annual Resource Recovery Conference, March 28–29 in Washington, DC. Washington: National Resource Recovery Association; 1985.

7 Energy and materials markets

Chapter Outline

7.1 Introduction

The primary objective of any proposed WTE project is to dispose of solid waste in an economically efficient and environmentally acceptable manner. Although the derived energy from the combustion of municipal solid waste in a WTE project is a secondary benefit, the revenues gained through the sale of electricity, steam, or recovered materials help offset the cost of solid waste disposal [1]. By turning solid waste into useful energy and materials, these projects are able to compete against other waste disposal options that are much less capital intensive. The financial viability of any WTE facility, therefore, is largely dependent upon its ability to sell energy to a long-term customer [2]. This chapter will explore some of the issues project developers should consider in investigating the energy and materials markets for any WTE project.

7.2 Energy markets

7.2.1 Steam

Prior to the Arab Oil Embargo in 1973, most incinerators in the United States recovered little or no thermal energy, except for internal plant needs. The worldwide increase in energy costs after 1973 made energy recovery from solid waste a viable

Waste-to-Energy. DOI: 10.1016/B978-1-4377-7871-7.10007-3

option. Many American cities that had once established steam distribution systems for their downtown areas, but abandoned them in the era of cheap energy, began to consider the redevelopment of these systems seriously. The steam produced by incinerating solid waste was considered as an energy source for steam distribution networks of single or multiple customers.

Typically, steam from WTE facilities in the United States is transported to industries at temperatures ranging from 250 to 750°F (120 to 400°C) and pressures ranging from 150 to 650 pounds per square inch. Pipelines transporting steam are short, rarely longer than two to three miles, in order to minimize energy losses and capital investment. When steam is transported it experiences a drop in pressure and temperature. Steam losses from such systems are estimated to be about 30 to 40 pounds per square inch per mile of transmission line. Consequently, the location of the energy customer is an important factor in the siting of a WTE facility.

Steam is the most desirable form of energy from solid waste, if an appropriate long-term steam customer can be located. However, the revenues which the developer of a WTE facility will receive from any customer are dependent on the price of the customer's existing energy service; the customer's steam needs, on a continuous basis; and the guaranteed reliability of the steam produced by the WTE facility. The highest prices for steam are received by those facilities that can continuously produce steam on a schedule that matches the customer'. Most facilities are unable to produce a guaranteed steam supply, and thus produce a limited steam supply requiring the customer to have a backup steam source. Consequently, the value of this steam to the customer is lower.

In summary, a WTE developer should consider the following critical factors in the marketing of steam:

- Siting – The facility should be close in proximity to the steam customer(s);
- Price – The steam produced must be delivered at a price competitive with the customer's existing primary source. Typically, the steam is sold in the form of hot water, per Btu, mostly for district or industrial building heating. WTE facilities can produce hot water in a less costly manner than steam, with low-grade energy or recovered heat, keeping a higher simultaneous power production. Heat sold in the form of hot water is typically the form of energy that is the most profitable for WTE plants;
- Operating schedule – Steam must be produced at a level matching the needs of the customer;
- Availability of fuel – Enough waste must be available to produce the contracted quantity of steam;
- Cost of connecting facilities – Who should pay the cost of installing steam lines, acquiring right-of-ways, and installing condensate return lines between the facility and the steam customer(s);
- Steam quality – Who is responsible to keep contaminants out of the steam lines and meet the individual customer's steam quality requirements; and
- Contracts – Can the steam customer sign a long-term put-or-pay contract; what will happen if they default.

7.2.2 Electricity

A long-term steam customer for a WTE facility may be unavailable in many parts of the world. In those instances, the conversion of solid waste into electricity may be a feasible alternative. The electricity generated by the facility can be utilized for internal needs of the project with surplus power being available for sale to utilities; transferred (wheeled) over the transmission of one utility to another; or made available for use by other governmental operations [3]. In some parts of the world – due to local regulations and subsidies – a WTE plant will sell its entire power production, but buy electricity for its own use under a separate contract.

Prior to the enactment of the Public Utilities Regulatory Policies Act of 1978 in the United States, commonly known as PURPA, WTE facilities seeking to sell their surplus power to public or investor-owned utilities faced significant obstacles. For example, many utilities refused to purchase electricity from such producers or offered to purchase power at low rates. In addition, some utilities charged discriminatorily high rates for back-up electric service.

In order to conserve oil and natural gas, the US Congress enacted PURPA with the objective of removing obstacles for facilities that generated electricity from alternative fuels such as solid waste, wood waste, and biomass. Under Section 201 of PURPA, electric utilities are required to provide electric service to 'qualifying facilities' at rates which are 'just and reasonable, in the public interest, and which do not discriminate against cogenerators and small power producers.' Qualifying facilities (QFs) are defined in the Act under two categories: small power production facilities, and cogeneration facilities. A small power production facility is one that produces less than 80 megawatts of electricity from biomass, waste, or renewable resource fuel constituting more than 50 per cent of the total energy source. Non-renewable fuel sources, such as oil, natural gas, or coal, cannot constitute more than 25 per cent of the total energy input of the QF during any year. In addition, an electric utility or its subsidiary may not hold more than 50 per cent of the equity interest of a QF.

Cogeneration facilities are somewhat different from small power producers in that they can produce and sell electricity and other forms of energy such as hot water or steam. Topping-cycle facilities are those which produce electric power first and then thermal energy. Bottoming-cycle facilities are those which produce thermal energy first and then electric energy. For such facilities to be deemed as QFs, PURPA requires that they meet certain specific operating and efficiency standards, and the same ownership criteria as small power producers.

Certification as a qualifying facility may take two forms under PURPA. The cogenerator or small power producer can merely 'notify' the Federal Energy Regulatory Commission (FERC), which administers PURPA, that their facility is a QF. This option made sense for those facilities whose characteristics do not vary dramatically from those prescribed under PURPA. The other option available for those projects is to submit an application to FERC requesting the agency to certify the qualifying status of the facility. In December 1985, FERC instituted a filing fee of US$7,800 for processing such certifications.

Under administrative rules promulgated by FERC to implement PURPA, utilities must purchase electric energy and generating capacity made available by QFs. The electric power purchased must be at rates reflecting the cost that the purchasing utility can avoid to produce or purchase from other utilities as a result of obtaining energy and capacity from the QF. Hence the terms 'avoided energy' and 'avoided capacity.' The FERC rules require that utilities must furnish data on avoided energy and capacity so that QFs are able to estimate such revenues.

While the FERC rules set the general policy requiring utilities to purchase electricity from QFs, the state public utility/service commissions are charged with enforcing this right. As one might imagine, each state commission has implemented the FERC rules quite differently in light of conditions existing in its own state. Consequently, the electric revenues and the type of permitted contractual relationship between utilities and a QF vary from state to state. For example, some states like New York had established a statutory six cents per kilowatt hour (KWh) floor payment by electric utilities to cogenerators and small power producers, while others like California have established complex formulas for computing avoided cost buyback rates. Several states have even allowed QFs to receive higher avoided capacity payments than utilities would incur in building new generating capacity because the power generated by cogenerators and small power producers is seen as more reliable than that from central generating stations.

Negotiating power sales contracts has been both a complex and expensive process for developers of WTE projects. Such contracts often form an integral part of the financing of these projects since electric revenues constitute a major portion of their long-term revenue bases. In response to the need to streamline the negotiating process between utilities and QFs, the increasing trend for state public service commissions has been to produce 'standard contract offers,' establishing a tariff which all electric utilities in their state must offer to virtually any qualifying facility. These standard contracts include provisions on subjects such as: regulatory risk; security requirements for early payments by utilities; production reliability; and interconnection requirements. For example, qualifying facilities in Florida, which execute a 'Standard Offer Contract' can receive avoided energy and early capacity payments equal to those costs incurred by a hypothetical 'Statewide Avoided Unit.' QFs are required by the Florida Public Service Commission, however, to secure these early capacity payments through a letter of credit, escrow account, or similar cash equivalent security, or in the case of a governmental project, by its full-faith and credit or other similar security.

Since 2005, the QF status under PURPA is no longer valid for all WTE plants. FERC rule 292.309 waives the obligation of utilities to purchase power from QF if the QF has nondiscriminatory access to wholesale markets and transmission and interconnection services. WTE plants bigger than 20 MW, and located in areas where access to a Regional Transmission System (RTO) or Independent System Operator (ISO) (such as MISO, PJM, ISO New England, NYISO, Cal ISO, ERCOT, SPP) is available, are deemed to have nondiscriminatory access to wholesale markets and transmission and interconnection services. In these instances, WTE plants bigger than 20 MW are not eligible to sell power to local utilities under QF status. However, if a plant bigger than 20 MW can prove that transmission constraints keep it from having access to markets, or if it is located in an area where no direct access is possible to an RTO/ISO, then it is eligible to sell power under the

QF status. WTE plants smaller than 20 MW are eligible to sell power under QF status.

Other possibilities for WTE to sell power in the USA are:

- Direct sale to industrial customers. This is not possible at the distribution level in regulated states, as local distribution companies have the monopoly to provide power on their service territory. In those states, only auto-consumption is possible (the owner of the power plant must be the same as the power customer).
- Direct sale to industrial customers through a transmission system. It is legally possible in most places but hard to implement because being hooked up to transmission system as a power market participant is very challenging for small plants.
- Sale to power markets need to be a 'market participant' (see above);
- Negotiated purchase agreement with a utility, at a negotiated rate or at the 'as available rate.'

Sales of carbon credits (for new plants) and/or renewable energy credits are possible (in some states, WTE is considered to be a renewable form of energy, in some states only the organic portion is renewable, and in other states WTE is not considered renewable), at least on the voluntary market. Legislation is evolving quickly on those issues and latest rules should be checked upon initiation of a new project.

7.3 Materials markets

The recovery of reusable materials from the solid waste stream has had a long history in the United States. These materials have traditionally been recycled from the waste stream, particularly paper and metals, when market conditions are favorable. Unfortunately, however, at times when markets for recovered materials are poor, due to fluctuations in the business cycle, these materials are unrecovered and must be disposed of by local government in sanitary landfills and WTE facilities. Generally not recognized by WTE project developers is the fact that materials recovery programs are compatible with the implementation of WTE projects and complement each other. For example, given the abrasive nature of glass, its removal would be a welcome relief for processing plant maintenance crews. Some of the advantages that a viable materials recovery program can offer include:

- Recovering glass and metals from the waste stream will provide a higher quality fuel for a WTE facility in most cases;
- Recycling removes materials which might be damaging to a WTE system;
- Materials recovery reduces the quantity of materials which would ordinarily be landfilled (ash residue, and non-processable wastes) in a WTE facility;
- A viable materials recovery system can help reduce the overall size and costs of a WTE facility;
- Conserves essential natural resources, including energy; and
- Helps extend the life of landfills.

The following subsections will discuss the types of materials that can be recovered and recycled from municipal solid waste; the technologies available for recovering these materials; and the types of markets typically available for such materials.

Materials in the municipal solid waste stream generally targeted for recycling include the following:

- Ferrous metals;
- Non-ferrous metals;
- Glass;
- Paper products; and
- Film plastics.

7.3.1 Ferrous metals

Ferrous metals, which are available in the solid waste streams of most communities, are primarily in the form of tin-coated steel cans and white goods. These metallic materials represent about 5 to 10 per cent by weight of the municipal waste stream in many US communities. They can be sorted out manually in source separation programs, or by magnets through front-end separation processes in RDF systems which can typically recover 80 to 85 per cent of ferrous metals from the incoming waste. Mass-burn systems can also employ permanent or electromagnetic separators and trommels to remove ferrous metals from the ash residue, but the incineration process produces a slag containing contaminants. Thus, these ferrous materials have significantly less value to scrap dealers, and the prices quoted are substantially lower than if such materials were recovered before incineration.

The markets and price received for ferrous metals vary with the existing industry demand, the quality of the material recovered, and the cost of transporting these materials to the market. Many WTE projects recovering ferrous metals from the waste stream have had difficulty negotiating long-term contracts due to the cyclical nature of the ferrous market.

7.3.2 Non-ferrous metals

Aluminum is the only non-ferrous metal that can be economically recovered from municipal solid waste. It usually represents less than 1 per cent by weight of the typical municipal solid waste stream. Other non-ferrous metals, such as lead and copper, are usually recycled by industries producing these materials.

The aluminum industry purchases aluminum from municipal source separation programs in the form of discarded beverage containers that are reasonably clean, dry and free of glass, steel, paper, and other contaminants. This market continues to be strong due to the high cost of producing aluminum cans. High-tech reverse vending machines placed outdoors in supermarket parking lots or lobbies are a recent trend.

Aluminum can also be recovered from the solid waste stream through either manual or mechanical means. Manual separation is highly labor-intensive, but results in the recovery of about 45 per cent of the aluminum in the waste stream. A number of mechanical and media separation techniques use trommels, disc screens, eddy currents, and other devices, depending on the differences in the specific gravity of aluminum. Experience to date with the latter equipment has not been entirely successful.

7.3.3 Glass

Municipal solid waste contains an average of 4 to 8 per cent glass by weight. Glass can be recovered from the municipal solid waste stream, either as broken or non-broken glass, and also sorted into different colors of glass such as flint (clear), amber, or green. Color separation enhances the market value of the glass material. Mixed glass material, termed 'cullet', has had poor marketability due to the glass industry's need for glass materials that are clean and free from refractory particles for the manufacture of clear bottles.

In recent years, the glass industry has increased its interest in recovered glass materials. Many manufacturers now recycle their own internally generated cullet. The use of such waste glass in their production processes allows the glass industry to reduce their fuel consumption since cullet is easier to melt than virgin raw materials. In order to maintain a desired minimum cullet percentage in their bottling facilities, they are also increasingly becoming buyers of cullet in the open market.

7.3.4 Paper products

Waste paper products, such as newsprint, corrugated paper (cardboard), and high-grade office paper (computer printout paper, tab cards, and ledger paper), typically represent about 30 to 40 per cent by weight of the municipal solid waste stream. These materials must be processed through source separation programs since mixing them in the solid waste stream renders them unacceptable for recycling. Separation of newsprint, corrugated paper or boxboard, and high-grade office paper has proven to be economically feasible in many areas, although the prices for these materials have fluctuated dramatically depending on market conditions.

7.3.5 Film plastics

The average municipal solid waste stream contains about 1 to 4 per cent plastic materials. The percentage of plastic materials in the waste stream has increased dramatically because of the surge in its use for a variety of household products, particularly in the field of plastic film packaging. Unfortunately, recycling of these products is relatively nonexistent. Some facilities in Europe are utilizing mechanical processes to recover the film plastics from the waste stream and manufacture plastic products such as trash bags and pipes.

7.4 Projected energy production from a proposed WTE facility

The energy generated from the incineration of solid wastes can be recovered as thermal energy, which is released from the flue gases exiting the combustion unit. Typically, the flue gases in WTE units are projected to range from 1,598 to 1,796°F (870 to 980°C) passing through the waste heat boiler downstream of the secondary

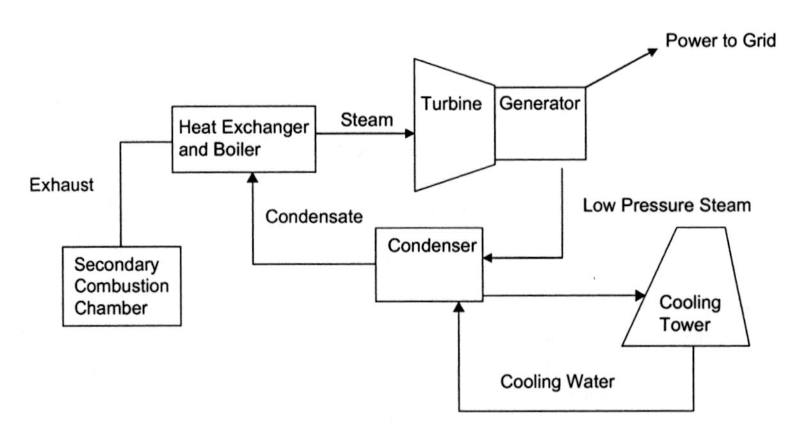

Figure 7.1 Traditional Rankine Cycle system.
Source: SCS Engineers.

combustion chamber. Boilers come in many different shapes, sizes, and designs. However, almost all follow the Rankine Cycle system, as noted in Figure 7.1.

To calculate potential levels of energy production from a proposed WTE facility, it is necessary to make general assumptions regarding the heating value of the waste, thermal efficiency of the boiler, efficiency of the turbine generator, energy requirements of the air-pollution control system, and any other in-place energy uses. A variety of turbine types and boiler designs can be used for the proposed WTE facility with ramifications on overall capital costs for the facility and energy production efficiency.

On top of plant size and fuel heating value, the design choices having the greatest impact on power production efficiency are:

- The pressure of the steam produced by boilers (standard at 40b/600 psi or high pressure at 65b/950 psi or higher);
- The type of condensing device (air cooled or water cooled);
- The heat source to pre-heat combustion air (if pre-heating is necessary): steam or energy recovered on flue gases;
- The number of boilers and turbines.

References

[1] Scaramelli Alfred B. Energy Market: Key to project planning. Solid Wastes Management 1982. April.

[2] US Environmental Protection Agency. Guides for Municipal Officials: Markets. Washington: US Environmental Protection Agency; 1976. SW-157.3.

[3] Davitian Harry. Power sales contracts for resource recovery projects. In: Proceedings of the Fifth Annual Resource Recovery Conference, National Resource Recovery Association; Washington, DC, March 28–30, 1986.

8 Permitting issues

Chapter Outline

Waste-to-Energy. DOI: 10.1016/B978-1-4377-7871-7.10008-5

8.1 Introduction

Long before construction activities for a WTE facility can begin, project participants must have received all the necessary regulatory approvals for construction and operation. The environmental permitting process can potentially be the most time-consuming and controversial step on the road to implementation for many WTE projects. This is due in part to the extensive data needed on such projects that must often be submitted to governmental agencies in the form of detailed permit applications and environmental impact statements. This chapter provides a general overview of the types of major environmental permits most likely to be required for a proposed WTE facility.

8.2 US solid waste combustor air emissions

Early discussions with environmental regulators usually reveal that a major environmental regulatory issue for any WTE project must deal with a review of potential air emissions from the project [1]. The following sections briefly describe the regulatory process in the United States for applicants requesting WTE operating permits under the Clean Air Act (CAA).

Section 129 of the CAA authorizes the US Environmental Protection Agency (EPA) to issue emission regulations for a myriad of emission sources, including solid waste combustors. Emission regulations specifically directed at the larger solid waste combustors were issued during the 1980s and 1990s, and defined the pollutants to be regulated and the limits of discharges of those pollutants into the atmosphere. The regulations also prescribed which control technologies would be used. The EPA has subsequently promulgated regulations addressing concerns about emissions from smaller solid waste combustion units. This section discusses air emissions and air emission control technologies consistent with the solid waste combustor air emission regulations issued by the EPA.

8.2.1 Air emission characteristics

There are three main categories of emissions from municipal solid waste combustors (MSWCs): acid gases; particulate matter (PM); and products of incomplete combustion (PICs). Within these three categories, there are numerous products of combustion, many of which are regulated under the CAA emissions regulations titled, 'New Source Performance Standards' (NSPS) for new sources and 'Emission

Guidelines' (EGs) for existing facilities. In this chapter, the discussions are limited to the following major municipal solid waste combustor air emissions:

- PM;
- Cadmium (Cd), mercury (Hg), and lead (Pb);
- Hydrochloric acid (HCl) and sulfur oxides (SO);
- Nitrogen oxides (NO_x);
- Carbon monoxide (CO); and
- Organics (dioxins and furans).

The daily, monthly, and yearly emissions concentrations from solid waste combustors vary. However, in general, the relative concentration and toxicity of any of the above contaminants in solid waste combustor gas will be affected by the following factors:

- Solid waste composition;
- Combustion temperature and residence time;
- Flow patterns and amounts of excess air; and
- Furnace design.

Section 129 to the CAA identifies five categories of solid waste incineration units:

- Units with a capacity of greater than 250 tpd combusting municipal waste ('large units');
- Units with a capacity equal to 35 tpd or less than 250 tpd combusting municipal waste ('small units');
- Units combusting hospital, medical, and infectious waste;
- Units combusting commercial or industrial waste; and
- Unspecified 'other categories of solid waste incineration units' that combust less than 35 tpd ('very small units').

In 1997, EPA established final standards for the performance of all existing and new municipal solid waste combustors [2]. NSPS limit the air emissions from facilities constructed, modified, or reconstructed after December 20, 1989, with a design capacity of 225 Mg/day (250 tpd) or greater. Existing MSWCs with design capacities of 225 Mg or greater are regulated under the Emission Guidelines.

The Emission Guidelines apply to all facilities with a design capacity greater than 250 tpd that were constructed, modified, or reconstructed prior to December 20, 1989. EPA divides existing MSWCs into two categories, with different emissions criteria for each category. Large MSWCs are defined as facilities with a design capacity between 250 and 1,100 tpd. Very large MSWCs are defined as those with a design capacity greater than 1,100 tpd. Under these guidelines, states (including US protectorates) are required to develop standards that are equal to, or more stringent than, the EPA standards that specify MSWC operating parameters, including CO levels, combustor load levels, and flue gas temperatures.

The 1997 NSPS regulate municipal solid waste combustor air emissions, which are comprised of organics, acid gases and metals, and nitrogen oxide (NO). The emission limits in the NSPS are based on the best-demonstrated technology (BDT) for solid waste combustors at their regulated capacity. The BDT for reducing emissions includes good combustion practices (GCP), a spray dryer, and fabric filter. The BDT for reducing NO_x emissions is an emission limit based on application of selective non-catalytic reduction at a typical facility.

In December 2005, EPA promulgated performance standards to reduce emissions from the last remaining category of waste incinerators requiring CAA regulation. The category is called 'other solid waste incinerators,' which consists of very small municipal waste combustion units that burn less than 35 tpd of MSW. Table 8.1 shows the current emission limits for new large, small, and very small MWCs, as promulgated by EPA.

8.2.1.1 Particulate matter

PM is any liquid or solid that is so finely divided it is capable of becoming windblown or suspended in air or gas. Particle sizes for PM from solid waste combustors usually range from 0.01 to 300 microns (micron, one-millionth of a meter) in diameter; 20 to 40 per cent of PM is less than 10 microns in diameter; and up to 10 per cent is less than 2 microns in diameter. PM from solid waste combustors is usually composed of the following materials: carbon particles, water particles, and particles of incomplete combustion.

The 1971 NSPS under the CAA brought about the first pollutant (particulate matter) control requirements for municipal solid waste combustors (the US Congress Air Pollution Act, 1955). Currently there are two principal technologies for the control of PM: electrostatic precipitators (ESPs); and fabric filters. Figure 8.1 illustrates typical electrostatic precipitators together with a detailed illustration of the charged plates within an ESP. It illustrates a fabric filter together with detail of the fabric bags within a fabric filter. Both ESPs and fabric filters are discussed in more detail below.

8.2.1.2 Heavy metals

At the end of the 1990s, three heavy metals emissions were regulated, mercury (Hg), lead (Pb), and cadmium (Cd).

Although heavy metals are considered to be in their own class of toxic emission, they are sometimes grouped with particulates. This may be attributed to the manner in which they are collected.

8.2.1.3 Acid gases

Sulfur dioxide (SO_2) and hydrogen chloride (HCl) are acid gases formed during combustion. Sulfur oxide (SO_x) compounds are formed form the oxidation of elemental sulfur, and HCl is formed when chlorine in the solid waste combines with free hydrogen atoms.

Spray dryers, sometimes called 'dry scrubbers,' are the most common acid gas control technology (Figure 8.1). Spray dryers inject a fine alkaline slurry (i.e., lime) into the flue gas stream. This slurry reacts with SO_x and HCl to form chlorine and sulfur salt precipitates, which are then collected by the electrostatic precipitators or fabric filters. The EPA regulations identify fabric filters as the best control technology

Table 8.1 Current EPA emission limits for large, small, and very small MWCs

Criteria	Large MWCs	Small MWCs	Very small MWCs
Carbon monoxide:			
Fluidized bed, mixed fuel	200 ppmdv	40 ppmdv*	
Mass-burn rotary refractory	100 ppmdv		
Mass-burn rotary water wall	100 ppmdv		
Mass-burn water wall and refractory	100 ppmdv		
Mixed fuel-fired (pulverized coal/refuse-derived fuel)	150 ppmdv		
Modular starved-air and excess air	50 ppmdv		
Spreader stoker, mixed fuel-fired	150 ppmdv		
Stoker, refuse-derived fuel	150 ppmdv		
Dioxins/furans	13 ng/dscm	13 ng/dscm	33 ng/dscm
Particulate matter (PM)	20 gr/dscf	24 mg/dscf	0.013 gr/dscf
Opacity	10%	10%	
SO_2	30 ppmdv or 80% Control	30 ppmdv or 80% Control	3.1 ppmdv
HCl	25 ppmdv	25 ppmdv or 95% Reduction	15 ppmdv
NO_x	180 ppmdv	500 ppmdv	103 ppmdv
Hg	50 μg/dscm	0.080 mg/dscm or 85% reduction	74 μg/dscm
Pb	140 μg/dscm or 150 ppm after first year of operation	0.20 mg/dscm	226 μg/dscm
Cd	10 μg/dscm	0.020 mg/dscm	18 μg/dscm

All emission limits (except opacity) are measured at 7 per cent oxygen, dry basis at standard conditions.
μg/dscm = micrograms per dry standard cubic meter.
ppmdv = parts per million dry volume.
gr/dscf = grains per dry standard cubic foot.
ng/dscm = nanograms per dry standard cubic foot.
mg/dscm = milligrams per dry standard cubic foot.
*12-hour rolling average using a continuous monitoring system.
Source: Reference [6].

Figure 8.1 Typical fabric filter installation and fabric filter.

(BCT) to combine with spray dryers for controlling acid gas emissions [1]. The use of a spray dryer with a fabric filter can produce removal efficiencies of >95 per cent for HCl and >85 per cent for SO_x.

8.2.1.4 Products of incomplete combustion (PICs)

PICs are of concern because carbon monoxide (CO) and organics such as dioxin or furan form by incomplete combustion. CO and organics emissions are regulated by the EPA.

Most organics and CO are destroyed when complete combustion occurs. This destruction requires ample amounts of oxygen. The EPA regulation of PICs from solid waste combustors has brought about the injection of excess oxygen to ensure that complete combustion will occur. In some cases, as much as 200 per cent of what would be stoichiometrically required for complete combustion, is added. Even with high excess air, CO and organics are still formed. Most of the PICs combine with particles in the cooling flue gas produced by dry sprayers; the particles are then collected by electrostatic precipitators or fabric filters. This process alone can produce removal efficiencies as high as 99 per cent for organics.

8.2.1.5 Nitrogen oxides

Nitrogen oxides are produced during combustion. The two most important nitrogen oxides are NO and nitrogen dioxide (NO_2). Collectively known as NO_x, these are formed from the organic nitrogen present in solid waste and from the nitrogen present in the air used for combustion. NO emissions are of concern due to their contributions to the formation of ozone and the photochemical oxidants known as smog.

8.2.2 Compliance monitoring

The USEPA monitoring under the NSPS and EGs requires major sources of air pollution (municipal solid waste combustors are considered to be a major source) to provide enhanced air emissions monitoring in order to demonstrate compliance with the CAA. Compliance monitoring is discussed below.

8.2.2.1 Oxygen

Municipal solid waste combustors must conduct continuous emissions monitoring (CEM) to measure oxygen at each location within a municipal solid waste combustor where CO, SO_x, or NO_x emissions are observed.

8.2.2.2 Particulate matter and opacity

The initial compliance tests for PM emissions and opacity mud must be conducted within 60 days after a municipal solid waste combustor has achieved maximum operating capacity, but not later than 180 days after startup. Following initial compliance testing, performance tests for PM and opacity are typically conducted annually.

Samples for PM compliance are collected from the air emissions in the municipal solid waste combustor stack. Sampling equipment (sampling trains) with probes and filters traverse the interior diameter of the stack to collect samples of the air emissions leaving the stack. Temperature conditions, sampling sites, the number of traverses, and the length of the sampling runs are dictated by EPA regulations. Other pollutant samples, e.g. metals and HCl, can be collected during the same sampling runs.

Municipal solid waste combustors are required to install a continuous opacity monitoring system (COMS) to measure opacity. Initial compliance with this EPA requirement must occur within 180 days of startup. Opacity is defined as the apparent obscuration of an observer's vision to a degree equal to the apparent obscuration of smoke given on the Ringlemann Chart. The Ringlemann Chart is a well-established standard for measuring opacity and photographically reproduces illustrations of four shades of gray that an observer can use to estimate the density of smoke. A clear stack is recorded as '0' and a 100 per cent black smoke as '5'.

8.2.2.3 Metals

The initial tests for compliance with emission limits for cadmium (Cd), lead (Pb), and mercury (Hg) must be conducted within 60 days after a municipal solid waste combustor has achieved maximum operating capacity, but no later than 180 days after startup. Samples to determine compliance of these metals are collected from the air emissions in the municipal solid waste combustor stack (see the section above on particulate matter and opacity for stack sampling details).

8.2.2.4 Sulfur dioxide

Municipal solid waste combustors are also required to install a GEM system to measure SO_2 emissions. Compliance with SO_2 emission limits is determined by calculating a 24-hour daily geometric mean emission concentration. At a minimum, valid paired CEM hourly averages (i.e., SO_2 and O_2) have to be obtained for 75 per cent of the hours per day for 90 per cent of the days per calendar quarter of operation. The initial performance test must be completed within 180 days after startup.

8.2.2.5 Nitrogen oxides

Municipal solid waste combustors are required to install a CEM system to measure NO_x emissions. Compliance with NO_x emission limits is determined by calculating a 24-hour daily geometric mean emission concentration. At a minimum, valid paired CEM hourly emissions (NO_x and O_2) have to be obtained for 75 per cent of the hours per day, for 90 per cent of the days per calendar quarter of operations.

8.2.2.6 Hydrogen chloride

The initial tests for compliance with HCl emission limits must be conducted within 60 days after a municipal solid waste combustor has achieved maximum operating capacity, but no later than 180 days after startup. Following initial compliance testing, HCl compliance is determined by annual stack tests.

Samples for HCl compliance are collected from the air emissions in the municipal solid waste combustor stack (see section 8.2.2.2 on particulate matter and opacity for stack sampling details).

8.2.2.7 Dioxin/furans

The initial tests for compliance with emission limits for dioxin and furans must be conducted within 60 days after achieving maximum operating capacity, but no later than 180 days after startup. After initial compliance testing, dioxin/furan compliance is determined by annual stack tests. Compliance is based on either the total emitted or the Toxic Equivalency Factor for dioxins and furans.

8.2.3 Operating standards

The EPA's NSPS and EG requirements also demand municipal solid waste combustors to demonstrate that the operations are conducted at rated capacity and that the emissions are in compliance with the tested and rated capacity of the facility. The EPA has established operating standards for a number of operational characteristics to ensure that the facility is operating at the tested and rated capacity. These standards address:

- Carbon monoxide – The amount of carbon monoxide in the emission gases is an indication of the degree of combustion; high carbon monoxide levels indicate insufficient air for complete combustion;
- Load level – A municipal solid waste combustor is tested for emission compliance at a rated capacity; defined by the steam load delivered. Therefore, the load level of steam being delivered is an indication of how well the facility is meeting the rated capacity and its emission level limits;

- Particulate matter control device temperature – For maximum efficiency, PM control devices must operate at their appropriate temperatures. Measuring the temperature at the inlet of the device demonstrates that it is ready to provide the intended PM removal.
- Municipal solid waste unit capacity – Each combustion train (unit) is rated at a certain capacity during compliance testing; measuring the 24-hour operating capacity demonstrates satisfaction of this requirement.
- Fly ash and bottom ash fugitive emissions – Municipal solid waste combustors must include means to prevent fugitive emissions from their ash handling system. Measuring the extent of fugitive emissions during the transfer of ash from a combustion unit to the ash storage facility determines compliance with this requirement.

8.2.4 Air emissions control technologies

This section presents an overview of the principal air emission control technologies used with municipal solid waste combustors.

8.2.4.1 Electrostatic precipitators

Electrostatic precipitators (ESPs) have very high removal efficiency rates and work quite well in high temperature flue gas that can impair the effectiveness of many control devices. ESPs use positive and negative electrical charges to collect particles. As the gas enters the ESP, it passes through an electrical field that charges the entrained particles. The particles are then collected on a series of large, oppositely-charged plates called fields. ESPs usually have several fields, and generally more fields result in higher removal efficiency. The plates must be cleaned periodically by 'rapping' them to remove the collected particles. If the plates are not rapped often enough, the removal efficiency will decline, and the plates may even become a source of particulates. However, with proper maintenance, ESPs can maintain removal efficiencies of approximately 99 per cent or more.

Many years of acceptable operating experience have proven the use of ESPs to control particulate emissions from municipal incinerators. When compared with other control technologies, ESPs prove to be the most cost effective, reliable, and most frequently applied technology for particulate control from major combustion sources, including resource recovery facilities. Several factors affect the collection efficiency of an ESP system. These include:

- Collection plate surface area per unit gas flow rate (specific collection area);
- Gas velocity through collectors;
- Number, width and length of gas passages;
- Electrical field strength and number of fields;
- Particle in field residence time; and
- Particle size distribution and fly ash resistivity.

Altering the design characteristics of the above parameters attains the desired control efficiency. The technical advantages of an ESP are:

- High collection efficiency and reliability;
- Relatively low energy consumption;

- Wide range of flue gas temperature applicability; and
- Minimal fire hazard potential.

The potential operational problems encountered during MSW combustion affecting ESP performance are:

- Corrosive condensation of flue gas constituents;
- Variation in particulate concentration in the flue gas;
- Reduced efficiency for very small particulates (less than one micron diameter); and
- Decreased collection efficiency caused by increasing particle resistivity, decreasing moisture content, and increasing flue gas-flow rate.

The primary technical disadvantage of ESPs is the poor collection efficiency for small particles. This deficiency can be resolved by increasing the specific collection area or by increasing the number of fields, but these changes increase capital and operating costs.

Preventing cold spots where condensation can occur can alleviate corrosion problems. This is accomplished by avoiding air leakage into the ESP and by insulation or external heating of the unit. Corrosivity is normally not a problem as long as the flue gas temperature is maintained above the sulfuric acid dew point temperature of 220 to 230°F (104 to 110°C). Fly ash resistivity is more predictable, and particulate collection efficiency increases in this temperature range.

8.2.4.2 Fabric filters

Because of increasingly more stringent regulations for PM, fabric filters, or baghouses, have become the particulate control technology of choice in the United States. Fabric filters act similarly to vacuum cleaner bags, i.e., as the gas passes through them, and the filter captures the particulates.

Tubular fabric media (bags) contained in multiple modular units is used to filter the exhaust gas. Initial particulate collection forms a thick porous cake of particulate on the bags. This cake increases the collection efficiency with the bag serving as structural support for the cake. As the cake builds, the pressure drop across the fabric filter increases, and eventually the cake must be removed. Fabric filters are categorized by the methods used for cleaning the bags: shaker, reverse gas, or pulse jet. The shaker method is not suitable for mass burn facilities without acid gas controls because the fabric required for this high temperature application cannot structurally withstand the violent shaking action over an extended operating period. The two other systems clean the bags by reversing the gas flow through the bags. The reverse gas system uses an external centrifugal fan with cleaned flue gas, and the pulse jet system uses compressed air. Figure 8.1 illustrates a typical fabric filter and a schematic of a fabric filter.

Some advantages of the fabric filtration systems are:

- High control efficiency for smaller particle size, including condensable organics and trace metals;
- Collection efficiency independent of flue gas composition and particle concentration; and
- Insensitivity to electrical resistivity of the fly ash.

Some disadvantages of fabric filters are:

- Corrosion that will occur if acid gas controls are not applied;
- Susceptibility to fires without flue gas quenching;
- Loss of bag structural integrity at temperatures above 550°F (288°C);
- Cementation, binding, or clogging of the filter fabric in a humid low temperature gas stream (assumes quenching or acid gas controls); and
- Relatively limited operating experience in WTE facility applications.

Fabric designs are currently available that minimize the effects of temperature, corrosivity, abrasion, and ignitability. Bag life can be increased when the fabric and cleaning methods are properly selected. Gas velocity (air-cloth the use of acid gas controls ahead of the fabric filter can effectively eliminate the potential problems associated with corrosivity, abrasion, and ignitability. Furthermore, a semi-dry scrubber can be operated to avoid clogging of the filter fabric. Operating experience in Japan, Europe, and at several plants in the United States using fabric filters in conjunction with acid gas controls has been satisfactory.

8.2.4.3 Venturi-web scrubbers

Venturi-wet scrubbers spray water across the gas flow to 'wash out' particles. They are less efficient than both ESPs and fabric filters and this is not considered as a good stand-alone control technology.

8.2.4.4 Wet scrubbers

Wet scrubbers control acid gases. Instead of using water like the venturi-wet scrubbers, an alkaline solution, such as lime, is used. The solution reacts with the acid gases to form salts, which are then collected as a sludge, dewatered, and landfilled. The most commonly used acid gas controls are dry sorbent injection (DSI) and spray dryer systems. DSI systems inject a dry alkali sorbent into the flue gas to reduce emissions of HCl and SO. In a spray dryer (also called a semi-dry scrubber or wet/dry scrubber), a lime slurry is injected into a drying chamber.

8.2.4.5 Catalytic reduction and selective non-catalytic reduction

Catalytic reduction and selective non-catalytic reduction are used to control NO_X and are more complex technologies than ESPs and fabric filters. In both procedures, aqueous ammonia or urea is injected into the furnace or flue gases to break down the NO gases into N. The main method used in the United States is selective non-catalytic reduction (SNCR). SNCR has demonstrated reductions of about 50 per cent. Selective catalytic reduction (SCR) is commonly used in Europe and Japan and has demonstrated removal efficiencies averaging about 80 per cent. Figure 8.2 shows an installation with this technology illustrating the ammonia injection; Figure 8.3 shows the ammonia tanks needed for this type of equipment.

Figure 8.2 Installation using ammonia injection.

Figure 8.3 Installation illustrating use of ammonia tanks.

8.2.4.6 Spray dryers/dry scrubbers

Spray dryers/dry scrubbers are used to control heavy metals. Spray dryers inject an alkaline slurry into the gas stream, causing water in the slurry to evaporate resulting in a cooling of the combustion gases. This cooling essentially de-vaporizes the heavy metals, allowing them to condense onto particulates. An electrostatic precipitator or fabric filter can then easily capture the heavy metal-bearing particulates. The USEPA recommends the use of a fabric filter for the most efficient removal of heavy metals. This method of collection usually provides removal efficiencies of about 99 per cent. However, the removal efficiency for mercury is only about 50 per cent. This lower efficiency is a cause of concern due to the high toxicity of mercury.

A semi-dry system (or spray dryer) uses a wet reagent lime slurry injection. Spray dryers remove acid gas in a two-stage system that consists of a spray dryer absorber (SDA) followed by particulate control. A SDA uses either a dual fluid nozzle or a rotary atomizer injection system to create fine droplets of reagent slurry. Small droplets have a greater surface area to volume ratio and by increasing the surface area contacting the flue gas, absorption of acid gases in enhanced.

The reagent slurry preparation area includes a reagent (lime) storage silo, reagent slakers, a slurry storage tank, and a dilution water tank. The dry reagent is mixed with dilution water in the slakers to form a slurry, which is stored in the slurry tanks. The injection rates of slurry and dilution water are controlled by signals from the in-stack CEMs and the SDA outlet temperature.

The combustion gases leave the economizer section of each boiler and pass through a dedicated SDA. The reagent slurry injected into the flue gas reacts with the acid constituents, neutralizing them and forming alkaline salt particles (Figure 8.4). The moist slurry also quenches any incandescent particles before entering the fabric filter. The slurry moisture evaporates from the heat of the combustion gases, thereby controlling the inlet temperature to the baghouse.

Superstoichiometric ratios of slurry are maintained in the SDA to ensure excess alkalinity, and as the precipitate accumulates on the surface of the fabric bags, additional acid gas removal is achieved. Fabric filters used in conjunction with a spray dryer scrubbing system have not been susceptible to failure on WTE facilities. While this application of fabric filters is relatively new, it appears that many of the disadvantages of using fabric filters alone are eliminated or substantially reduced. These combined systems are not susceptible to fires due to the long retention time in the spray dryer and the cooling of the gases. High temperature excursions are also not likely and the addition of the alkaline reagent results in more efficient filtering characteristics and lower corrosion potential.

Some advantages of the semi-dry scrubber/fabric filter system are:

- Simultaneous removal of acid gas and particulate;
- Reduced corrosion potential allowing wider latitude in materials of construction, reducing capital cost. Maintenance costs are also reduced;
- Equipment life extended;
- Flue gas temperature above the dew point resulting in no visible condensation plume;
- Flue gas reheating not generally required for plume rise;
- High removal efficiencies for PM, including very fine particles and trace metals; and
- Dry stable residue for less costly disposal.

Some disadvantages of the semi-dry scrubber system are:

- Potential for clogging and erosion of the injection nozzles due to the abrasive characteristics of the slurry;
- Potential for slurry or particle build upon walls of the SDA and ductwork;

Figure 8.4 Dry scrubber atomizers.

- Potential for blinded bags if slurry moisture does not evaporate completely; and
- High reagent usage since reagent is not recycled.

Figure 8.4 illustrates the use of atomizers for injection of the reagent slurry.

8.2.4.7 Dry reagent injection/fabric filtration

In a dry reagent injection/fabric system, fine powdered alkaline reagent (lime or soda ash) is continuously injected into the flue gas duct upstream of a fabric filter. A bulking agent is added to reduce the pressure drop through the fabric filter, thereby increasing the time between cleaning cycles. The reagent and the bulking agent build up a caked layer on the filter fabric that then serves as a reactor bed to neutralize acid gases. Early assessments of this unproven technology indicate that its principal advantages are:

- Lower equipment costs than with other systems because there is no slurry preparation, slurry pumps, or reactor vessel;
- Lower operating costs than other systems because there are no internal moving parts; and
- Lower power consumption.

The disadvantages of the dry reagent injection system are:

- That it is prone to caking and potential plugging of injection system due to moisture absorption;
- The powdered reagent is more costly than the pebble form used by most semi-dry systems; and
- Higher stoichiometric ratios than spray dryers are required.

Wet scrubbing systems contain three basic components for the actual gas-scrubbing portion of the system:

- An inlet spray quench chamber to cool the gases to the adiabatic (without gain or loss of heat) saturation temperature with either fresh water or a scrubbing reagent;

- A contacting section where the gas stream and scrubbing liquor are in intimate contact to promote absorption and neutralization of the acid gases by the scrubbing reagent; and
- An entrainment or mist eliminator section to remove liquid droplets entrained in the gas system leaving the contacting section to prevent particulate emissions from dissolved or suspended solids after evaporation of water from the scrubbing liquid droplets.

Wet scrubbing systems may operate with either sodium- or calcium-based reagents, although sodium reagents are more costly than calcium reagents. A wet scrubbing system functions by dispersing the scrubbing liquor in the gas stream to create a large surface area for absorption. The various gas components removed from the flue gas stream are absorbed by the scrubbing liquor, and neutralized by chemical reaction with the scrubbing reagent. Reaction products and unreacted reagent are removed from the system for disposal. For a lime system, the by-product is a wet sludge consisting of the calcium salts of chlorine, fluorine and sulfate, excess reagent, and particulates removed in the scrubber. The moist scrubber gas is exhausted through the system stack.

Some advantages of wet scrubbers are:

- High acid gas control efficiencies;
- Long operating history and proven performance in acid gas removal applications, although not with WTE facilities; and
- Control efficiency that is relatively independent of temperature and humidity of the inlet gas.

Some disadvantages of wet scrubbers are:

- Materials of construction must be corrosion resistant, hence costly;
- Flue gas will be saturated resulting in a visible condensation plume;
- Relatively poor plume dispersion due to low temperatures, unless stack gas reheat is used, which consumes energy and increases capital and operating and maintenance costs;
- High energy consumption per unit of pollutant reduced;
- The scrubber liquor must be treated to enable liquor recycling and avoid excess materials cost; and
- Wastewater (blowdown) discharged from the scrubber water system will frequently require treatment before discharge to sewers (sewer ordinances generally specify the maximum allowable concentration of total dissolved solids, and an acidic stream may contain unacceptable levels of heavy metals in solution).

Some wet scrubbers have been used outside the United States for particulates and HCl removal. However, in the US, they have not been demonstrated in commercial operation on large solid waste fired boilers.

8.2.4.8 Activated carbon

Many WTE facilities have installed a powdered activated carbon injection system to control mercury (Hg) emissions. Mercury is a metal that exists in the liquid phase and is relatively volatile at ambient temperatures. As a product of combustion, the quantity of Hg emitted is a function of the Hg content of the solid waste burned in the furnace. Mercury and mercuric compounds have boiling points ranging from 575 to 1,200°F (300 to 650°C). These compounds vaporize when incinerated due to their relatively low vapor pressures and can have a tendency to remain in the vapor phase at the outlet of the air pollution control equipment.

One theory is that flue gas cooling encourages deposition or enrichment of these compounds on fine PM.

Activated carbon or a modified activated carbon material may be injected upstream of the spray dryer or in the spray dryer to achieve Hg reductions. According to some activated carbon technology manufacturers, powdery activated carbon can be continuously fed into the flue gas upstream or downstream of the spray dryer absorber. The contact between Hg and carbon in the filter cake of the downstream fabric filter provides the necessary reaction time.

Carbon injection may achieve significant reduction of Hg emissions, as shown by two sets of patent test data. In three pilot test runs, injection of 80 mg/Nm^3 in flue gas at a location between the SDA and fabric filter resulted in overall Hg reductions of 89 per cent (230°F run [110°C run]), 95 per cent (230°F run [110°C run]), and 91 per cent (284°F run [140°C run]). The Hg reductions with no carbon injection were about 69 per cent. Carbon injection can also remove as much as 90 per cent of dioxin.

8.2.4.9 Combustion controls

Emissions of CO can be reduced at solid waste combustion facilities by using good combustion practice (GCP) to maximize the oxidation of CO to CO_2. GCP includes proper design, construction, operation, and maintenance. Important factors in facility design and operation that reduce CO emissions include:

- Maintaining uniform solid waste feed rates and conditions;
- The use of preheated air to burn wet or difficult-to-combust materials; maintaining adequate combustor temperature and residence time;
- Providing proper total combustion (excess) air levels;
- Supplying proper amounts and distribution of primary (underfire) and secondary (overfire) air;
- Minimizing PM carryover;
- Monitoring the degree of solid waste burnout; and
- The use of auxiliary fuel during startup and shutdown.

GCP reduces CO emissions as well as organic emissions by promoting more thorough combustion of these pollutants.

Several features can be incorporated into a WTE facility to promote GCP and to minimize emissions of CO. The grate and furnace design can incorporate multiple, individually controlled air distribution chambers below the grate structure and grate to promote proper air/fuel mixing in the primary combustion zone. Similarly, the secondary (overfire) air system can be designed to induce turbulence, provide adequate flame length control, and prevent the stratification of unburned gases.

In addition to combustion air control, furnace flue gas temperatures should be maintained at 1,800°F (980°C) for at least one second. These temperatures, combined with proper supply of overfire air, will result in maximum oxidation of the CO and other hydrocarbons resulting in minimum CO emissions.

8.3 International air emission regulations

8.3.1 European Union

Similarly to the United States, there was strong political will in Europe during the 1980s to enhance existing environmental regulations. There were initial attempts to implement air emission standards in Europe such as the EU Directives on prevention/reduction of air pollution from new/existing waste incinerators [3] and German TA Luft [1], which required implementation of 'best available technologies' (i.e., electrostatic precipitators, fabric filters, and combustion control) to clean up the flue gases from existing and new facilities.

These directives were soon followed by additional legislative measures during the 1990s to further ratchet up the regulatory standards, as new technologies (scrubbers, activated carbon, $DeNO_X$, etc.) were proven reliable in actual plant operations. As a consequence, the European Union (EU) promulgated more stringent air emission regulations to prevent or reduce impacts on waste incineration or co-incineration on human health. In order to limit these risks, the EU's Directive in December 2000 [4], and a further modification in 2005, impose strict emission limits for certain potential pollutants released to air and water [5].

Briefly, the Directive requires that all plants (incineration and co-incineration) must maintain flue gases at a temperature of at least 1,262°F (850°C) for at least two seconds. For those incinerating hazardous wastes with content more than 1 per cent halogenated organics (expressed as chlorine) the temperature must be raised to more than 2,012°F (1,100°C) for at least two seconds. The operators of waste incinerators are also required to monitor emissions to ensure that they comply, as minimum, with the limits in the Directive and country air regulations.

Table 8.2 shows the limits for different air emissions, as noted in Annex V of the Directive. This directive mandates the strictest emission limit values for WTE as compared to any other industry in Europe. Today, the EU standards for WTE are copied worldwide as the best environmental protection standards for WTE facilities (Table 8.3).

8.4 Solid waste combustor ash management

For many solid waste agencies in the United States, solid waste combustors are an important part of their integrated solid waste management system. The per cent volume reduction of solid waste, which is achieved through combustion, enables these agencies to extend the life and capacity of their existing landfills. Since ash residue produced as a result of solid waste combustion is typically 25 to 30 per cent by weight (10% by volume) of the incoming solid waste tonnage, one of the key issues surrounding the development and implementation of any solid waste combustor is ash management. In this section, the generation of municipal solid waste combustor ash, the characteristics of ash, ash regulation in the United States, and the management of ash are discussed [6].

Table 8.2 EU air emission limits

Parameter	Europe waste incineration directive Emission limits for plant capacity > 6 t/hr (mg/m^3 @ 11% O_2)	Typical plant of 1000 t/d Max allowable emission rate (kg/ton of waste incinerated)
		101.3
SO_x	50	0.233
HCl	10	0.0466
NO_x	200	0.932
CO	50	0.233
HF	1	0.00466
Dust	10	0.0466
CO_2	None	73.21

Note: A new directive (IED) has been approved by the European Parliament in 2010 and will be finally approved by the European Council in the coming months. The latest version of the new IED directive will keep the same emission limits as in the table above but change the units to dry mg/Nm^3 at 11% O_2.
Source: Deltaway Energy, 2010.

8.4.1 Generation

Ash from a municipal solid waste combustor is composed of two ash streams: fly ash – the finely sized, airborne particles of ash entrained in the flue gases and collected by air pollution control devices; and bottom ash – the non-airborne combustion ash discharged at the end of the grates.

The characteristics of these two ash streams are significantly different. Fly ash contains products of incomplete combustion including acid gases, vaporized and volatilized heavy metals, and many other complex organic and inorganic compounds. Bottom ash is essentially inert material discharged from the grates. Ash management practices in the United States combine these two ash streams within the facility to form municipal solid waste combustor ash. Ash, like any other solid waste, is tested to determine whether it is a hazardous waste. Test results from a number of municipal solid waste combustor units throughout the United States during the early 1990s indicated that this solid waste does not normally test as hazardous. In Tables 8.4 and 8.5 are listed the many inorganic and organic constituents that can exist in ash from municipal solid waste combustors.

The common methods of managing solid waste combustor ash are either utilization or landfilling. Ash can be utilized for a number of useful purposes including for road-building materials and in the manufacture of Portland cement. The metals from the bottom ash can also be recovered for recycling. Ultimately, whether some utilization is practiced or not, some portion of the ash will have to be landfilled. Landfilling is either achieved by co-disposal with municipal solid waste or other solid wastes or mono-filled in specially dedicated landfill cells. The most common method of landfilling combustor ash in the United States is by co-disposal with solid waste.

Table 8.3 Comparison of United States, European, and Asian emission standards

Concentration units		Pollutant	France	Europe	USA	Thailand	Taiwan	China	Macau
ppm	7% O_2	SO_x (ppm at 7% O_2)	26.7	26.7	30	30	102.02	127.7	98.24
ppm	7% O_2	HCl (ppm at 7% O_2)	9.4	9.4	25	25	51.01	64.7	86.3
ppm	7% O_2	NO_x (ppm at 7% O_2)	136.8	136.8	150	180	229.54	273.5	342
ppm	7% O_2	CO (ppm at 7% O_2)	61.1	61.1	100	188.8	153.03	168.5	112.35
mg/m^3	7% O_2	HF (mg/m^3 at 7% O_2)	1.4	1.4	8	8	8	8	7.02
mg/m^3	7% O_2	Particulate matter/dust (mg/m^3 at 7% O_2)	14.06	14	24	120	281	112.3	140.4
TEQ ng/m^3	11% O_2	Dioxins (TEQ ng/m^3)	0.1	0.1			0.1	1	1.2
$\mu g/m^3$	7% O_2	Cadmium ($\mu g/m^3$ at 7% O_2)	OC	70.2	20	444	380	140.4	OC
$\mu g/m^3$	7% O_2	Mercury ($\mu g/m^3$ at 7% O_2)	70	70.2	80	444	380	280.8	OC

Source: Deltaway Energy, 2010.

Table 8.4 Ranges of concentrations of inorganic constituents in ash from MWC incinerators (ppm)

Parameter	Fly ash	Combined ash	Bottom ash
Arsenic	15–75	2.9–50	1.3–24.6
Barium	88–9,000	79–2,700	47–2,000
Cadmium	<5–2,210	0.18–100	1.1–46
Chromium	21–1,900	12–1,500	13–520
Lead	200–26,600	31–36,600	110–5,000
Mercury	0.9–35	0.05–17.5	ND–1.9
Selenium	0.48–15.6	0.10–50	ND–2.5
Silver	ND–700	0.05–93.4	ND–38
Aluminum	5,300–176,000	5,000–60,000	5,400–53,400
Antimony	139–760	<120–<260	–
Beryllium	ND–<4	ND–2.4	ND–<0.44
Bismuth	36–<100	–	ND
Boron	35–5,654	24–174	85
Bromine	21–250	–	–
Calcium	13,960–270,000	4,100–85,000	5,900–69,500
Cesium	2,100–12,000	–	–
Cobalt	2.3–1,670	1.7–91	3–62
Copper	187–2,380	40–5,900	80–10,700
Iron	900–87,000	690–133,500	1,000–133,500
Lithium	7.9–34	6.9–133,500	7–19
Magnesium	2,150–21,000	700–16,000	880–10,100
Manganese	171–8,500	14–3,130	50–3,100
Molybdenum	9.2–700	2.4–290	29
Nickel	9.9–1,966	12–12,910	9–226
Phosphorus	2,900	290–5,000	3,400–17,800
Potassium	11,000–65,800	290–12,000	920–13,133
Silicon	1,783–266,000	–	1,333–188,300
Sodium	9,780–49,500	1,100–33,300	1,800–33,300
Strontium	98–1,100	12–640	81–240
Tin	300–12,500	13–380	40–800
Titanium	<50–42,000	1,000–28,000	3,067–11,400
Vanadium	22–166	13–150	53
Yttrium	2–380	0.55–8.3	–
Zinc	2,800–152,000	92–46,000	200–12,400
Gold	0.16–100	–	–
Chloride	1,160–11,200	–	–

ND = not detected at the detection limit.
Blank = not reported.
Source: Reference [6].

Table 8.5 Ranges of concentrations of organics

Constituent	Fly ash	Bottom ash
Naphthalene	270–9,300	570–580
Biphenyl	2–1,300	–
Acenaphthylene	ND–3,500	37–390
Anthracene	1–500	53
Fluorene	0–100	ND–150
Phenanthrene	21–7,600	500–540
Di-*n*-butyl phthalate	ND	360
Fluoranthene	0–6,500	110–230
Pyrene	0–5,400	150–220
Butyl benzyl phthalate	NDE	180
Chrysene	0–690	ND–37
Bis(2-ethylhexyl)phthalate	85	2,100
Benzanthrene	0–300	–
Benzo(k)fluoranthene	ND–470	ND–51
Benzo(a)pyrene	ND–400	ND–5
Benzo(g,h,i)perylene	0–190	ND
Diethyl phthalate	6,300	–
Acenaphthene	–	28
Normal alkanes	50,000	–
Chlorobenzenes	80–4,220	17
Chlorophenols	50.1–9,630	0

ND = not detected at the detection limit.
Source: Reference [6].

8.4.2 Regulation

Municipal solid waste combustor ash is not exempt from hazardous waste regulation under Section 3001(i) of the Resource Conservation and Recovery Act (RCRA). Like any other solid waste, it must be tested to determine if it meets the requirements of RCRA Subtitle C hazardous waste regulations.

In 1987, the Environmental Defense Fund (EDF) sued Wheelabrator Technologies and the City of Chicago for violating RCRA by not managing ash from their WTE as hazardous (*City of Chicago vs Environmental Defense Fund*). Due to uncertainties concerning the impact of federal ash regulations, studies of ash utilization and reuse have been limited. However, there are some promising projects currently in progress exploring the use of ash as landfill cover, aggregate in road construction and in concrete blocks, and material for artificial reefs.

8.4.2.1 Sampling and analysis of solid waste combustor ash

The USEPA has provided test procedures, such as the toxicity characteristic leaching procedure (TCLP), to determine whether a solid waste is hazardous [6]. The TCLP test is intended to simulate the leaching of solid wastes that is disposed of in a RCRA Subtitle D sanitary landfill. The test consists of passing the solid waste material through a sieve, adding acetic acid to maintain a specified acidic pH throughout the test, and agitating the mixture for 24 hours. The mixture is then filtered and the liquid phase is analyzed to determine if any of the specified concentrations for eight elements (arsenic, barium, cadmium, chromium, lead, mercury, selenium, silver), four pesticides, and two herbicides have been exceeded. If the extract contains one of the above substances in an amount equal to or exceeding the levels specified in the RCRA Subtitle C hazardous waste regulations, the solid waste is determined to be toxic and is classified as a hazardous waste. Based on knowledge of the characteristics of municipal solid waste, the two most likely elements that may be found in amounts exceeding the regulatory thresholds are lead and cadmium. None of the other substances has been found in quantities approaching the regulatory thresholds.

The results of an EPA test may vary considerably depending on whether the test was performed on municipal solid waste combustion fly ash, bottom ash, or combined ash. Test data for fly ash sometimes shows lead and cadmium levels in excess of regulatory thresholds. Bottom ash test data show those elements in quantities well below test thresholds. The generally accepted explanation for the difference is that the heavy metals in the solid waste being incinerated are volatilized into the combustion gases by the high furnace temperatures. Then, as the gases cool in the boiler and air pollution control equipment, the metals condense onto the fly ash, thereby producing the higher quantities of metals in the fly ash leachate.

A fundamental problem in testing ash is obtaining a representative sample of the ash. The residue from the incineration process is extremely variable in size and composition. Obtaining a sample of the ash that is representative of the entire waste stream is difficult. Furthermore, the testing procedure applied to the sample produces results that are also extremely variable. It is quite common for the statistical analysis of the laboratory test results on all samples to reveal that there has not been sufficient sampling. For many facilities, producing statistically accurate laboratory test data is cost-prohibitive.

8.4.3 Characteristics

Combustion will reduce the volume of incoming solid waste by 85 to 90 per cent. The combustor's efficiency will determine the actual volume reduction. Specific factors determining the efficiency include the combustion chamber temperature, the air-to-fuel ratio, the degree of mixing, and residence time in the combustion chamber. Volume reduction percentages are also affected by the specific composition and moisture content of the incoming solid waste. The weight reduction after incineration is approximately 70 to 75 per cent.

Fly ash comprises approximately 10 per cent of the combined ash stream and the bottom ash makes up the remaining 90 per cent. As mentioned earlier, the common practice in the United States is to combine fly and bottom ash inside the facility prior to handling and disposal.

8.4.3.1 Physical composition

Given current combustion efficiency and air pollution control requirements, the ash from a solid waste combustor of the late 1990s in the United States can be classified as 'well burned'. Chemically, ash is predominantly silicon; aluminum, iron, and calcium oxides; sulfate; and chloride ions. The minor constituents include organic and inorganic compounds.

The characteristics of ash from MSWCs will vary considerably due to many factors. Changes in composition of the ash can be caused by seasonal variations of solid waste, the addition of new types of solid waste from surrounding local municipalities (i.e., new sources), or the implementation of local recycling programs. In addition, any changes in facility operating conditions and facility design, such as the introduction of a dry scrubber, could substantially change the composition and characteristics of the ash. Heavy metal concentrations would also be affected by changes in manufacturing standards that might limit the use of metals in the production process.

Development and implementation of more efficient air emissions control equipment should not result in substantial changes in the metal concentration of ash, since current state-of-the-art facilities already achieve >99 per cent removal efficiency. Sorting or processing solid waste prior to incineration improves the consistency of the ash product and the performance of an MSWC. Front-end separation of metal objects and metal-containing solid waste products will reduce the concentration of metals in the ash, although the reduction will probably not be significant in the fly ash portion. The sources of heavy metals in the solid waste stream are fairly well known. Sources of lead include batteries, metal scrap, auto parts, solder in cans, pipes, and paint. Lead is also found in ink pigments used in magazines and newsprint. Cadmium is used for coating and plating metal equipment parts, electronics, and rechargeable batteries. Mercury is found in disposable batteries, calculators, thermometers, meters, gauges, and cameras. MSWC ash generally contains higher levels of lead and cadmium than of mercury.

Municipal solid waste combustor ash characteristics vary according to the specific air pollution control equipment employed. Data from facilities using a spray dryer type scrubber indicates that the addition of lime during the process will generally increase the pH and thereby reduce the leachability of metals in ash. However, if sufficient lime is added to raise the pH above 12, leaching of certain metals (e.g., lead) at an elevated rate can result. A dry injection scrubber system will generally add higher amounts of lime than with the spray dryer type scrubber. The addition of lime will result in an ash product that is more likely to exhibit pozzolanic properties (for a cement-like substance in the presence of moisture). Although most existing data on MSWC ash comes from facilities that do not employ scrubbers, it is reasonable to assume that future facilities constructed in the USA will use scrubbers for air pollution control.

8.4.3.2 Inorganic and organic constituents

The fly ash component usually contains much higher concentrations of all metals (except copper and iron), polychlorinated biphenols (PCBs), polychlorinated dibenzo-*p*-dioxins (PCDDs), and polychlorinated dibenzofurans (PCDFs) than the bottom ash. However, bottom ash usually contains higher concentrations of semivolatiles than does fly ash. Results from numerous studies have indicated that PCBs,

PCDDs, PCDFs, and semi-volatile compounds are relatively immobile in the environment. These studies have also shown that lead and cadmium have the highest potential for leaching of any metal constituents.

8.4.4 Ash management alternatives

Currently available ash management alternatives are shown in Figure 8.5. The most appropriate option for disposal or treatment and utilization will obviously vary from

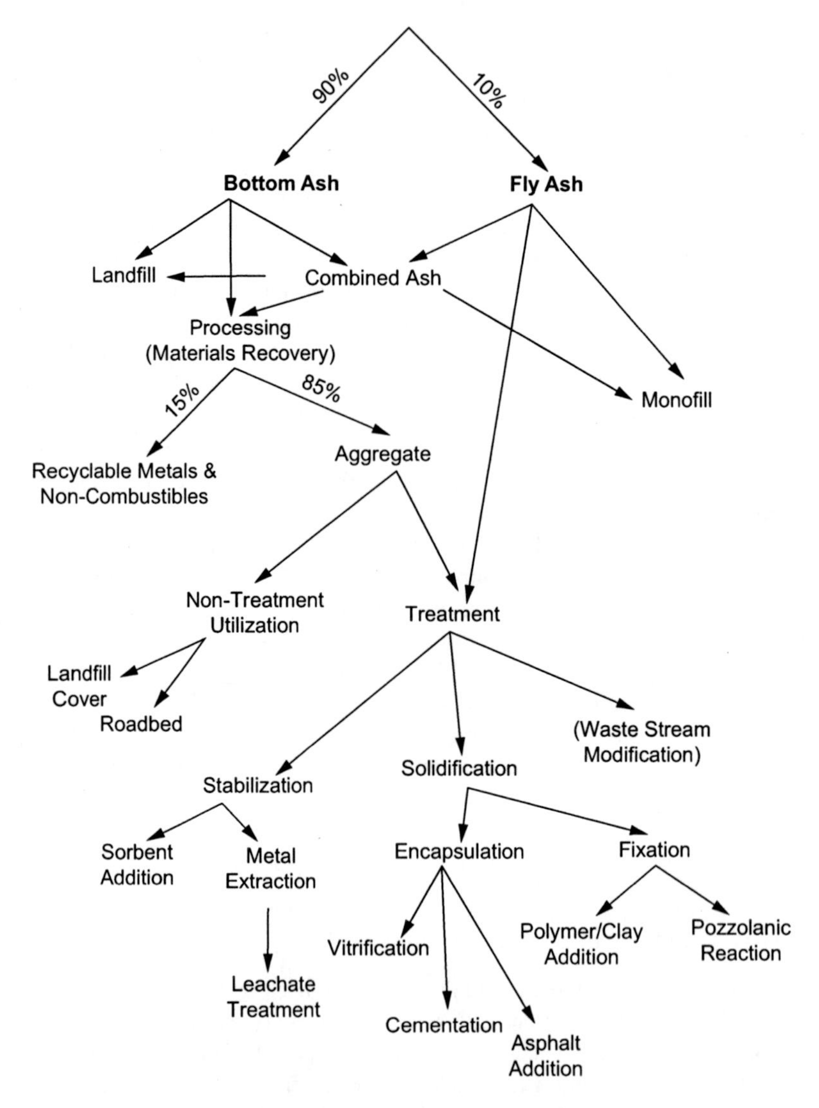

Figure 8.5 Municipal solid waste combustor ash management alternatives.

facility to facility depending on its geographical area, operational size, and specific ash characteristics. The selection of an appropriate ash management option is based upon several very important factors, which can be summarized as follows:

- The specific physical and chemical characteristics of the ash;
- The technology for treatment;
- The availability of potential markets;
- Local environmental effects of various disposal and reuse options; and
- Treatment or disposal costs.

Proper handling and transport of ash is important regardless of the selected management alternative. Ash should be thoroughly moistened prior to transport or handling in order to reduce dust emissions. Most facilities use internal quench tanks that result in an appropriate moisture content of the ash leaving the facility. Trucks used for transporting ash should be covered to avoid particulate loss during transport of the ash. Trucks should also be leak resistant if the ash will be transported off-site to prevent any leachate from escaping during transport.

8.4.4.1 Landfill disposal

Solid waste combustor ash may be landfilled in either mono-cells (cells dedicated specifically to ash) or co-disposed with either solid waste or hazardous waste. There are no specific US regulations regarding the landfilling of solid waste combustor ash. However, the ash must be tested to determine if it passes or fails the RCRA Subtitle C Hazardous Waste criteria. If it routinely passes the Hazardous Waste criteria, it is commonly considered as a non-hazardous RCRA Subtitle D solid waste.

The most common method of landfilling of solid waste combustor ash is by co-disposal with solid waste. Surveys in the late 1990s by the Solid Waste Association of North American indicated that 70 per cent of those solid waste combustors responding co-disposed of their ash with solid waste in regulated Subtitle D sanitary landfills [6].

8.4.4.2 Ash utilization

Ash utilization is common in the United States. Fly ash or bottom ash may be used with or without treatment. The technical issues associated with the use of ash are threefold: environmental considerations, equipment and system technology, and product marketability. The environmental considerations address such areas as siting, air emission impacts, and leachate production for each prospective use. There are essentially four basic benefits associated with ash utilization:

- Immobilization of metals;
- Cost-effective management;
- Conservation of natural resources; and
- Conservation of landfill capacity.

The first and most important goal is to immobilize the metals contained in the ash so that any utilization option is environmentally safe. The economic advantages are many, when the costs for landfilling of ash from solid waste combustors can range from US$20 to US$100 (1998 prices) per ton depending on the type of landfill used.

8.4.5 Utilization without treatment

In the USA, there were two alternatives that did not require treatment of the ash prior to utilization. These two methods were: use as landfill cover and as roadbed and fill material in road construction.

Where ash is used as landfill cover, various advantages have been realized. The ash product exhibits high workability and is spreadable and compactable even under wet conditions. It also provides the same benefits as any daily cover soil – control of blowing paper, fire control, and vector control. Studies conducted in Florida revealed that MSWC ash was much more weather resistant than natural soil as cover.

Screened bottom ash is used extensively in Europe as roadbed and fill material. Successful applications in the United States include studies conducted by the Federal Highway Administration and several cities and states.

8.4.6 Utilization with treatment

Because of the lack of information and data on treatment alternatives in the late 1990s, they are only described briefly here. The treatment of solid waste combustor ash falls into three broad categories:

- *Solidification* – solidification is considered a physical treatment;
- *Stabilization* – stabilization is considered a chemical treatment; and
- *Incoming solid waste stream modification.*

There are several types of solidification treatment:

- *Encapsulation* – This involves physical binding of the ash particles by adding materials such as asphalt or Portland cement;
- *Vitrification* – This is a type of encapsulation process which involves fusion of each particle in a kiln heated to approximately 2,600°F (1,425°C) resulting in various glass products, such as pellet aggregate, glass grit, and foam glass. The glass products provide an impermeable matrix.
- *Fixation* – This solidification treatment involves physical binding of the ash by adding materials such as polymers or clay. Fixation can also be achieved by enhancing the natural pozzolanic reactions in ash through appropriate management.

Stabilization is a chemical treatment process. There are two basic stabilization treatment alternatives:

- *Sorbent addition* – This process involves a chemical reaction between sorbent materials and trace metals in the ash residue to form insoluble compounds. Examples of sorbent materials include hydrous silicates such as clay and vermiculite. Clays are especially suitable for absorbing metal ions. The addition of sorbent material to ash residue causes a chemical reaction, stabilizing the waste product.
- *Metal extraction* – This process involves using an acid wash to extract heavy metals such as cadmium, lead, zinc, copper, silver, and gold. The 'leachate' is then treated by one of several

available processes to remove the metal components. Existing processes for treating the leachate include precipitation, ion exchange, and absorption. After metal extraction with the acid wash, the 'inert' residue can be used in a wide variety of ways.

As an ash stabilization treatment, metal extraction has not been investigated as extensively as other incoming solid waste, so stream modification may be misnamed as a treatment process. However, diversion, or reduction of certain materials that may be in the solid waste brought to an MSWC, can certainly alter the chemical characteristics of the ash. Examples of this would be banning lead acid batteries and mercury containing materials to reduce lead and mercury levels in the ash; and green wastes to reduce the amount of NO_X. Reduction of such materials, resulting in a more benign ash, would make disposal and utilization easier. To accomplish this, however, requires changing generation patterns and collection practices. This is not as easy to do as it sounds since it involves control of the generation, collection, and combustion process. In the free market, and no waste control conditions in integrated solid waste management in the United States in the 1990s, such control may not exist. Consequently, while this approach may make sense, external forces may not allow it to be done.

Sources of wastewaters from solid waste combustors include domestic wastewater, run-on/run-off from the surface areas of the property, cooling tower waters, boiler waters, air pollution control devices, ash management, and general facility cleanup waters.

Water is used in the combustion process for several purposes:

- Scrubber effluent for acid gas cleaning;
- Cooling and quench water from bottom ash management;
- Other air pollution control devices;
- Boiler feedwater production; and
- Cooling towers.

The wastewaters from the various uses can contain high concentrations of solids, heavy metals, and soluble contaminants.

Anticorrosion and antifouling agents are used at solid waste combustors in the boilers and in the cooling towers. They include:

- Corrosion inhibitors:
  - Boiler – hydrazine substitute (modified dehydroxy-1-amine)-oxygen scavenger;
  - Cooling tower – nonpolluting polysilicate/organic polymer-based corrosion inhibitors plus scale and foulant control.
- Chemical and biological antifouling agents:
  - Boiler – deposition and caustic corrosion control: sodium bi- and triphosphates;
  - Cooling tower – acrylic-based calcium phosphate inhibitor; and
  - Biocide – continuous treatment with chlorine is a routine procedure to maintain a minimum residual chlorine level of 0.1 ppm in the water supply basin.

Wastewater treatment methods include neutralization, removal of solids, precipitation, and settling. When possible, wastewaters are treated and recycled for additional use. Frequently, the management practice of choice is discharge into the sanitary sewer for treatment at the publicly owned treatment works (POTW) that is treating domestic wastewaters. Discharges to sanitary sewers are subject to the POTW's discharge limits.

References

[1] Eineitung und Eriauterungen von Ruth Henselder. Vorschriften zur Reinhaltung der Luft, TA Luft: Fassung vom 27. Germany: Bundesanzeiger; 1986.

[2] US Code of Federal Regulations (CFR), 40 CFR, Part 60:
Subpart Ea of 40 CFR Part 60 – New Source Performance Standards for Municipal Waste Combustors constructed after December 20, 1989 and on or before September 20, 1994.
Subpart Ca of 40 CFR Part 60 – **(withdrawn 1995)** Emission Guidelines and Compliance Times for Municipal Waste Combustors constructed on or before December 20, 1989.
Subpart Eb of 40 CFR Part 60 – New Source Performance Standards for Large Municipal Waste Combustors constructed after September 20, 1994.
Subpart Cb of 40 CFR Part 60 – Emission Guidelines and Compliance Times for Large Municipal Waste Combustors constructed on or before September 20, 1994.
Subpart FFF of 40 CFR Part 62 – Federal Plan Requirements for Large Municipal Waste Combustors constructed on or before September 20, 1994.

[3] EU Directives on prevention/reduction of air pollution from new/existing waste incinerators (1989). CELEX Nos 31989L0369 and 31989L0429.

[4] Directive 2000/76/EC. On the Incineration of Waste of the European Parliament and of the Council, December 4, 2000.

[5] European Suppliers of Waste to Energy Technology. Energy-From-Waste. Brussels: The Big Picture; October 11, 2010.

[6] Rogoff Marc. Current Developments in MWC Ash Management. Proceedings of the 28th Annual Solid Waste Exposition, GRCDA August 12, 1990.

9 Procurement of WTE systems

Chapter Outline

9.1 Introduction

The procurement of a WTE system by a community is one of the final steps on the long road of project implementation. Prior to embarking on this final path of system procurement, each community would have already addressed the difficult decisions, as discussed in previous chapters, which are necessary to implement the project [1]. The history of WTE project implementation has shown that many projects have suffered serious setbacks, if not abandonment, because their organizers had not

Waste-to-Energy. DOI: 10.1016/B978-1-4377-7871-7.10009-7

addressed these decisions. Oftentimes, these decisions were not made even after the procurement process had already begun.

Once these critical project decisions are settled, the procurement process for a WTE facility can begin. Communities have wide latitude in the approaches and procedures they can utilize in procuring WTE systems, although many statutes and local charters and ordinances may limit the choices available to some communities. Since a WTE facility is usually the most capital intensive and complex public works project attempted by the majority of communities, great care must be taken at the outset by a community's legal advisors to develop a procurement process which both meets the community's needs, and also follows state and local statutes [2].

This chapter will discuss some of the general approaches and procedures which may be used by local government to procure WTE systems. The procurement approach determines the manner by which engineering, design, construction, start-up, and operations services are acquired. The procurement procedure also dictates the method by which such services can be acquired under the legal mandates of the government.

A generic procurement process used will serve to illustrate how a community might develop a Request-for-Proposal (RFP), evaluate bidder responses, and conduct final contract negotiations.

9.2 Procurement approaches

The process selected by a community to procure a WTE system can be approached in three different ways:

- Design, Bid, Build (D-B-B) approach;
- Design, Build (D-B) approach; and
- Design, Build, Operate (D-B-O) approach.

Table 9.1 identifies some of the project responsibilities typically undertaken by government, contractors, and the A/Es involved in the WTE projects.

9.2.1 Design, Bid, Build (D-B-B) approach

The D-B-B approach is the one most generally used for large public benefit projects. For a typical project, the government agency or private owner first procures the services of an A/E firm, which is then responsible for preparing the plans or system specifications and certain design elements of the particular project. This material is then distributed by the community as part of an advertised competitive-bid process.

Once the winning bidder is selected, the A/E firm that prepared the project specifications and plans (or a similar type of firm) is oftentimes retained by the community to monitor the construction of the project, prepare operating manuals, and assist in the start-up and acceptance testing of the project. Once the community has

Table 9.1 Typical energy from waste project responsibilities by procurement approach

	Procurement approaches		
Project responsibilities	**D-B-B**	**D-B**	**D-B-O**
Planning	G or E	G or E	G or E
Preparation of plant specifications	E	C	C
Plant design	E	C	C
Construction supervision	E	C	C
Construction and equipment installation	C	C	C
Startup	G or C	G or C	C
Operation	G or C	G or C	C
Ownership	G	G	G or C

G, Government; E, Engineer; C, Contractor.
Source: Reference [5].

accepted the project as meeting contractual obligations, it is then responsible for either operating the project itself or for contracting out this responsibility to a private firm. This approach requires multiple contracts between the local government and its A/E, general construction contractor, and equipment suppliers.

In their procurement of WTE projects, some local governments or owners have slightly modified this conventional D-B-B approach. Unlike the traditional D-B-B approach of bidding each individual piece of equipment for the project, the entire process line and turbine generator equipment, commonly referred to as the 'chute-to-stack' in mass-burn facilities, is bid as a single package. The A/E firm is still responsible for designing the ancillary facilities, which can then be tendered out as individual bid packages, such as site civil and structural foundations and building construction to minimize contractor/vendor markups for profit and risk. This approach has the advantage of minimizing the number of potential vendors that the local government or private owner must deal with, while providing a mechanism for the government to share the risk of project performance with a private entity.

Typically, contractors responding to these design packages demand that their liabilities (warranties, liquidated damages) under construction contracts are, at the outside, capped to their own contract amount. That is, in the event the project is awarded through separate lots, any contractor's liabilities are capped in proportion with the value of each lot. Normally, contractors refuse liability for indirect damage (loss of profit, financial loss, repute damage, consequential damage, etc.). Therefore, the owner has to look for insurance coverage of such damage under this scenario. In this case, contractors generally accept the burden of insurance deductible. Further, within the limits of each contract value, contractors have to provide the owner with relevant sureties (surety bond, performance bond, warranty bond). Contractors generally require payment bonds from the owner, amounting to owner's maximum or average debt exposure.

9.2.2 Design, Build (D-B) approach

The D-B approach differs from the D-B-B approach in that a single private firm is responsible for the design, construction, and startup of the project. Under this approach, the turnkey contractor (Figure 9.1) is responsible for acquiring the necessary equipment and supplies for the project as well as ensuring that the architectural or engineering design work is prepared. Once startup and acceptance testing is completed, however, the turnkey contractor turns over the responsibility for operating the project to the community or owner. This approach may make perfect sense for those communities that have the technical capabilities to operate WTE facilities. Other communities or owners desiring to retain the option of operation may modify this approach by negotiating a short-term initial operating agreement, which would require the contractor to solve design and operational problems after acceptance testing before turning over full operation to a municipal workforce.

The typical turnkey agreement used in the WTE industry has the following important facets:

- The turnkey contractor is bound to the owner under a single contract that encompasses the whole of supplies and services relating to the works: design, construction, commissioning, and warranties;
- On account of this structure, the contractor takes upon any and all responsibilities and liabilities to the owner as engineer and project manager. In particular, the contractor is in charge of general and specific engineering, as well as the definition and management of interfaces, and the co-ordination of his sub-contractors;
- The contractor is fully responsible to the owner for the works he entrusts to sub-contractors;
- The contract liabilities apply to the overall contract amount, not to the split up value of each technical lot (or to sub-contracts) value. For example, if liquidated damages for lack of performance are limited to 15 per cent of the contract value, a default affecting the turbo-generator may give rise to a penalty amounting up to 15 per cent of the contract value, not 15 per cent of the turbo-generator value only (whereas if the contract is awarded in separate lots, liquidated damages for the defaulting turbo-generator supplier would be limited to 15 per cent of the turbo-generator value).

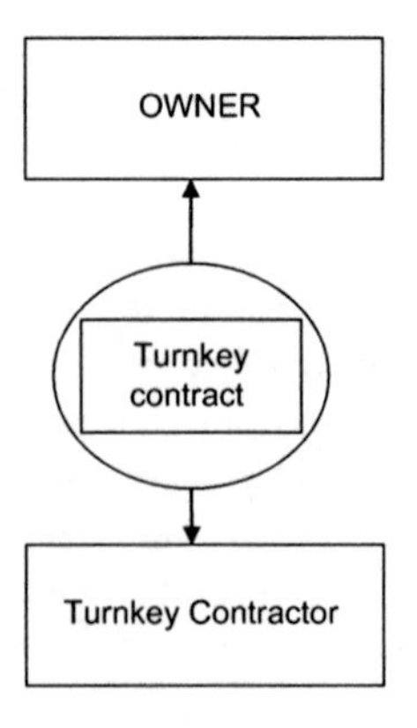

Figure 9.1 Relationship between owner and turnkey contractor.

Nowadays, contractors are more and more reluctant to enter into pure turnkey contracts, as they fear the prospect of being weighed down by liabilities for services out of their core business.

In order to obviate the difficulty to find pure turnkey contractors, owners may resort to 'partial turnkey' (Figure 9.2) contracts. Such schemes generally consist of the following components:

- One civil works contract (earthworks, outside piping, covering building and amenities, etc.);
- One equipment contract, including treatment process, flue gas cleaning, power generation, control–command, waste preparation and storage, etc.).

As shown in Figure 9.2, within each main lot the contractors are liable for their possible sub-contractors in a way similar to that described for the 'pure' turnkey contract. But the owner incurs the consequences of a default on these two contractors. The owner has to cope with general engineering, co-ordination, interface definition and management, which he or she may either perform directly, or sub-contract to external design-engineering specialists. Contractors are less reluctant to enter into such contracts, on account of reduced liabilities.

Typically, the types of contract drafted under this partial turnkey scheme include:

- Civil works contract;
- Process engineering, procurement and construction (EPC) contract;
- Engineer contract;
- Interface contract;
- Waste delivery;
- Power purchase;
- Slag tipping and slag recycling.

9.2.3 Design, Build, Operate (D-B-O) approach

A modification of the D-B approach is for a community to assign total responsibility to a private firm over the conduct of the project, including the design, construction,

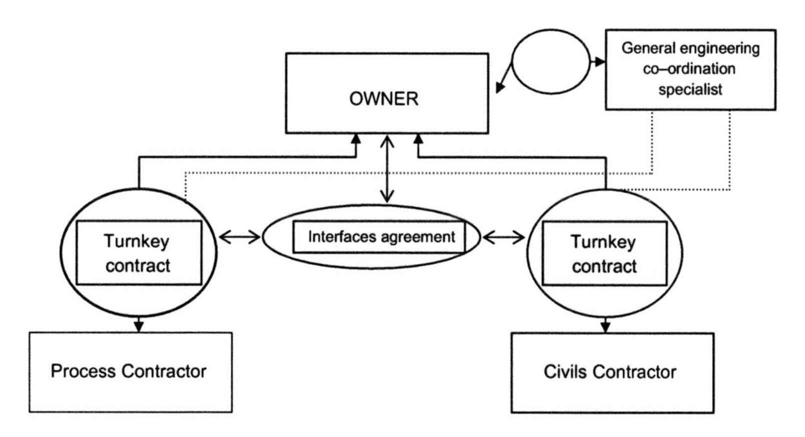

Figure 9.2 Partial turnkey contracts.
Source: Deltaway Energy, Inc.

startup, testing, operation, and possibly, the ownership and project financing responsibilities [3]. The full-service approach can enable a community or owner to acquire the services of a WTE facility without making the community responsible for its long-term, day-to-day operation and maintenance (O&M).

9.3 Procedures for conducting the procurement process

A basic tenet of procurement policy worldwide is that fair and open competition minimizes the costs of goods and services to a community. Competition is believed by many to improve the quality of the goods and services purchased, increase the choices available, and foster innovation among vendors. In addition to these goals, there is a general public belief that competition reduces favoritism and inspires greater confidence in the governmental procurement system.

WTE systems are highly complex and sophisticated public benefit projects. Procurement procedures that may be appropriate for obtaining public goods and services at the lowest possible cost for a traditional public benefit project may not be desirable in the case of a WTE facility. Discussed in the paragraphs below are four general procurement methods that the American Bar Association has recommended for projects in the United States [4]. Not all these procedures are applicable for use with the procurement approaches available for WTE facilities, as previously discussed. For example, the competitive sealed bidding procedure is not appropriate in turnkey or full-service procurement processes because factors like vendor experience, technology, guarantees, and financial capability, are considered along with price to select the winning contractor. Table 9.2 provides a summary of the relationship between procurement approaches and procedures.

9.3.1 Competitive sealed bidding

Competitive sealed bidding or formal advertising is the standard procurement contracting method used by most communities for acquiring equipment and services for public benefit projects [5]. Typically, the community or its consultant prepares the Invitation-for-Bid (IFB) used to solicit bids for the specific project. Contractors prepare their bids based upon the specifications described in this document with no

Table 9.2 Applicability of procurement approaches and procedures for WTE projects

	Procurement approaches		
Procurement procedures	**D-B-B**	**D-B**	**D-B-O**
Competitive sealed bidding	Appropriate	Not appropriate	Not appropriate
Competitive negotiation	Possible	Appropriate	Appropriate
Multiple step process	Not appropriate	Appropriate	Appropriate
Sole source	Not appropriate	Appropriate	Appropriate

Source: Reference [5].

discussions with the purchasing officials of the community. These sealed bids are usually opened in public at a time and place stated in the IFB. Generally, the bid is promptly awarded to the bidder submitting the lowest price, provided this bidder can satisfy the contractual standards of the community and is responsive with all the terms of the IFB.

Competitive bidding can be used in the procurement of a WTE facility, provided the following project conditions are present:

- Definitive specifications have already been prepared for the WTE facility;
- Price or some other factor is the only criterion of choice; and
- More than one response to the IFB can be expected.

9.3.2 Multiple-step or simultaneous negotiations method

This multiple-step procedure, sometimes called 'simultaneous negotiations,' has been utilized by the federal government for the procurement of advanced engineering and operational systems for which detailed specifications had not yet been prepared by the purchasing agency. These multiple-step procedures incorporate components of competitive bidding and competitive negotiation. At the outset of the procurement, an RFP is advertised, calling for the submittal of an unpriced technical proposal that will be evaluated by a selection committee upon receipt. Some governments have used a request-for-qualifications (RFQ) as a method to limit the number of proposers who receive the RFP. At this stage, discussions are held with proposers to determine the type of materials and services offered, and to allow proposers to make changes in their proposals to enable them to be responsive to the solicitation document. These discussions conclude with the offering agency requesting a 'best and final' offer on a common cutoff date by vendors. These final offers are then reviewed and evaluated. Thereafter, the process is similar to the competitive bidding process in that the proposer whose bid is most responsible and complete is awarded the overall contract.

In summary, the competitive bidding process is most useful when specifications are complete for a WTE project and price alone is the only factor for choosing the best overall vendor.

9.3.3 Competitive negotiation

Competitive negotiation is quite different from the competitive-sealed bidding process used for conventional public works projects [4]. In many communities, this process is reserved for the procurement of specialized services, such as architectural, planning, and engineering services, where discussions between the offering agency and vendor can help clarify the reasonableness of proposals. Consequently, competitive negotiation differs from the competitive sealed bidding process in two basic respects:

- Price is not the only criterion of choice. The use of other factors helps the offering agency to determine which proposal is most advantageous to the offering agency; and
- Discussions are held with vendors after the submittal of proposals to determine important technical, financial, and management interrelationships of each proposal.

Competitive negotiation has been the most common procedure utilized by communities to procure complex public works projects, such as WTE systems. Under this procedure, the community solicits proposals, not bids, with documents termed RFQ and RFP. In some cases, communities release a single procurement package, which contains both the RFQ and RFP, to speed up the procurement process. Typically, this procurement process can begin with the offering agency advertising the release of an RFQ to system vendors. This document contains some comments from the community on system performance, procurement schedule, desired technology, and financing requirements. In most cases, the RFQ requires vendors, who desire to prepare proposals on the project, to submit very specific technical, managerial, and financial qualifications regarding their undertaking of the project. For example, respondents to the RFQ may be asked to provide technical data on their performance capabilities, such as operational data from commercial-sized reference plants utilizing their technology.

Once these submittals are analyzed by the community, vendors deemed qualified to undertake the project receive a copy of the RFP package, which contains detailed performance specifications desired by the community for the WTE project. This document specifically describes criteria, which will be used by the community to evaluate the proposals. Shortly after the RFP is released, many communities schedule a pre-proposal conference to help brief the potential proposers on information contained within the RFP; provide clarifications to proposer's questions; and review the procurement schedule. If, as a result of this conference, and subsequent written questions submitted by proposers, changes in the RFP are determined to be advisable by the offering agency, amendments to the RFP can be released. Since the complexity of these changes may affect the responsiveness of all proposals, many communities have changed their deadline for proposal submission.

Following receipt of the proposals, and their detailed technical evaluation, the community may either begin discussions with a single top-ranked proposer, or enter into simultaneous negotiations with two or more of the top-ranked proposers. This option of simultaneous negotiations must be specifically described by the community in the RFP document. These discussions may cover any items contained within the proposals, as long as the information derived by one proposal is not disclosed to another competing proposer. Following these discussions, each proposer is offered the opportunity to revise its proposal and submit a 'best and final offer' to the community. After evaluation of these final offers, an award of the procurement is announced by the community.

9.3.4 Sole-source negotiation

A sole-source, negotiated procurement involves no competition, and is usually restricted to cases where the time factor for procurement is extremely critical or where there is only one source of supply for the desired WTE technology. Under a sole-source procurement, a vendor either submits an unsolicited proposal or one in response to an RFP. Upon receipt of this proposal, the community evaluates the proposal and negotiates the terms and conditions of the award of the project.

Sole-source procurement of WTE systems is often open to criticism based on favoritism and high costs due to the absence of market competition.

9.4 Preparing the request-for-proposals

The RFP document is critical to the success of a community's procurement process for a WTE facility. This document is designed to acquaint potential proposers with the specific technical, financial, institutional, and contractual aspects of the project. Clearly, it must be able to communicate the proposal requirements of the issuing agency to potential proposers. As such, the community must carefully draft this document to eliminate ambiguities that might result in some proposers not responding to the RFP. The RFP, therefore, is the community's statement of its goals for the project to the WTE industry [6]. Each RFP must be tailored specifically to the needs of the particular community. Consequently, a 'standard RFP' does not exist.

This section describes some of the information that an issuing agency should include in an RFP at a minimum for a WTE facility.

9.4.1 Format of a request-for-proposals

An RFP used typically has the following seven major divisions/sections [7]:

- Introductory materials – This section provides a listing of project team members and includes the table of contents;
- General information to proposers – This section provides proposers with an overview of the County and the project;
- Instructions for proposal preparation and submission – This section provides instructions for the preparation of proposals and how and when they should be submitted;
- Technical requirements – The purpose of this section is to inform proposers of the requirements, constraints, and technical conditions for the design, construction, and operation of the project;
- Proposal forms – This section provides proposal forms to be completed by each proposer;
- Draft agreements – This section provides draft design and construction and O&M agreements which require comment by each proposer; and
- Evaluation, selection and negotiation process – This section describes the process used by the County to evaluate each proposal, and how the negotiations process would be undertaken. It contains a listing of consultants and staff, such as engineers, investment bankers, and special counsels, who have responsibility in the format and preparation of the community's RFP. Such information is critical since the proposer's decision to respond to an RFP can be affected by the quality of the community's consulting team. Following this listing, a complete table of contents is provided for the reader's easy access to the document.

9.4.2 General information to proposers

This section is designed to provide proposers with an overview of the project. Introductory remarks usually indicate the community's desires to design, construct, startup, test, operate and maintain the WTE facility; furthermore, this introductory subsection indicates the location of the facility site, its permitting status, and the system size and generator configuration.

This subsection is followed by other subsections that provide background information on the community such as:

- Location, size and population;
- Governmental structure;
- Solid waste management system;
- Solid waste quantities;
- Energy market;
- Waste flow control;
- Permitting requirements and responsibilities;
- Site location and description;
- Subsurface conditions;
- Preliminary site plan;
- Utilities; and
- The financing plan including a description of the draft bond ordinance, tax issues, and bond validation.

9.4.3 Instructions for proposal preparation and submission

This section was designed to serve as a guide for preparing and submitting a proposal in response to the RFP. The explanation of the method of procurement is provided. These introductory remarks are followed by subsections providing general information on such topics as:

- Proposal submission location;
- Proposal submission deadline;
- Addendum and interpretations;
- Proposal preparation expense;
- Signature and authority requirements;
- Openness of procurement process;
- Security bond requirements;
- Errors and omissions by proposers;
- Retention and disposal of proposals;
- Schedule of project events subsequent to release of RFP;
- Pre-proposal conference; and
- Organization of proposal responses.

In some communities, purchasing requirements means that each copy of each proposal must be submitted in the following four physically separate and detachable volumes for evaluation purposes:

- Volume I: Executive summary;
- Volume II: Technical description;
- Volume III: Price proposal;
- Volume IV: Qualifications.

The remaining subsections provide detailed instructions for the proposers for preparation of each of these respective volumes. For example, the executive summary is limited to 30 pages, including tables, figures, and illustrations such as the artist's rendition of the Facility.

The technical proposal describes the major design and construction details of the project with specific instructions provided in the RFP on the following components:

- A project site plan;
- Process flow description and diagram;
- Project arrangement drawings;
- Major equipment description and specifications;
- Materials of construction;
- Process mass balance diagrams;
- Process energy balance diagrams;
- Process control diagram;
- Project availability analysis;
- An artist's rendition and landscaping plan;
- Performance guarantees;
- Electrical interconnection plan;
- Overall construction management plan project schedule;
- Project milestone schedule;
- Project startup procedures;
- Project operating plan;
- Personnel requirements;
- Utility utilization;
- Operational performance of project; and
- Environmental compliance.

The price proposal section for each proposal is designed to aid each contractor in summarizing prices and to assist the county in evaluating proposals. This subsection generally describes the proposal forms that each respondent must complete, as well as the instructions the proposer should follow in commenting on the draft design and construction, and operations and management (O&M) agreements included with the RFP. As part of the price proposal, the proposers are instructed to state all the conditions and terms in these draft agreements to which exception was taken, using legislative drafting format.

This instructional section concludes with a discussion of the required format of the qualifications proposal. Each proposer must provide specific details regarding its technical experience or its licensor's (if any) overall experience on similar projects, including actual records for such plants; detailed information on its full-time employees directly involved with the project; and detailed information on major subcontractors. Each proposer is instructed to provide financial data such as:

- Recent 10K filings with the US Securities and Exchange Commission in the United States or audited financial statements for the three previous years;
- The last quarterly financial reports for the prior two years;
- A full and complete description of its legal and financial guarantees;
- A description of its license agreement, if any, in terms of service and guarantees;
- A copy of any financial document offering any of the firm's financial offerings;
- Credit ratings; and
- A copy of each firms latest annual report, if any.

9.4.4 Technical requirements

The purpose of this section is to inform the proposers of the requirements, constraints, and conditions for the design, construction, and operation of the project. To facilitate evaluation of all technical proposals, the RFP can instruct the proposers to utilize design data and information presented with the RFP in the following areas:

- Project capacity. For example, a minimum of 372,000 tons per year of reference waste with a continuous design rating of 1,200 tons per day; and
- Reference waste composition, characteristics and ultimate analysis.

The purpose of this section is to specify the minimum design requirements for the project in the following areas of the project:

- Scale and weigh system;
- Tipping area;
- Solid waste storage area;
- Fire protection system;
- Fragmentizer;
- Overhead cranes;
- Combustion/steam generation units;
- Boiler feedwater and treatment system;
- Residue system;
- Air pollution control systems;
- Flues and stack;
- Power generation system;
- Electrical switchgear;
- Sitework, building and structures including architectural and landscape treatment; and
- Process control and monitoring system.

9.4.5 Proposal forms

This section includes various forms that the proposer must complete according to the instructions in the RFP and attach to the respective proposal volumes.

9.4.6 Draft agreements

This section includes the draft Design and Construction Agreement and Operations and Management Agreement that the proposer should comment on, using the legislative drafting format.

9.4.7 Evaluation, selection and negotiation process

This last section of the RFP describes the procedure the County would utilize to review and evaluate each proposal to first determine its completeness, and then to undertake the evaluation of each of the four volumes of each proposal. This section specifically describes the ranking and weighting process utilized according to some illustrative scores in the following areas:

- Technical evaluation: 20 per cent;
- Aesthetic and architectural appearance: 10 per cent;
- Qualification evaluation: 20 per cent;
- Economic evaluation: 30 per cent; and
- Agreements response evaluation: 20 per cent.

The section also describes the negotiations process the County should use once proposals have been evaluated.

9.5 Proposal evaluation

The process utilized by a community to evaluate proposals submitted in response to its RFP must be carefully developed to ensure that all proposals are reviewed and evaluated fairly. This evaluation process need not be time-consuming if proper pre-planning is undertaken to: select and train staff who will evaluate such proposals; develop the necessary comprehensive forms to help evaluators summarize data from each proposal; and develop and test computer models to analyze economic and financial data derived from the proposals. The following section describes a methodology typically used to evaluate responses to its RFP for a WTE system.

9.5.1 Log-in procedure and proposal handling

In accordance with instructions in the RFP, all sealed proposals can be received by the local government's Purchasing Department prior to the opening time prescribed in the RFP and addenda subsequently released. At the time of proposal opening, all proposals received by the community can be left unsealed, and the following general information is recorded for each: design and construction price, base operating fee per ton, excess operating fee per ton, and whether a bid and performance bond were included with each proposal, as required by the RFP.

Since the RFP may require numerous copies of all volumes of the entire proposal, one set is usually resealed and deposited in the community's 'Clerk's Office' for safekeeping. This is done in case questions regarding the accuracy of the remaining proposal sets arose later. The other nine sets of proposals were then transmitted to the County's Department of Solid Waste that was responsible for coordinating the proposal evaluation effort.

9.5.2 Evaluation committee

Prior to the release of the RFP, a proposal evaluation team for a WTE project should be composed of team members who are experienced in the areas of engineering, construction management, economics, finance, and law. They would include representatives from the following key departments: Office of the Administrator, Department of Solid Waste, Division of Public Works, Division of Fiscal Services,

Clerk's Office, and the Office of the Attorney. The following members of the County's consulting team would assist this panel: engineering consultant; financial advisor; insurance advisor; investment bankers; and legal advisors.

The following subcommittees are established, charged with the responsibility of reviewing and evaluating specific components of each proposal for ranking purposes:

- Technical evaluation;
- Aesthetic and architectural appearance;
- Qualifications evaluation;
- Economic evaluation; and
- Agreements response evaluation.

9.5.3 *Review of completeness and conformance with an RFP*

Following receipt of the proposals, they are reviewed by the subcommittees with respect to completeness, and in conformance with the instructions and requirements specifically indicated in the RFP. These requirements are mandated by official actions of the government agency.

9.5.4 *Detailed evaluation*

Each of the remaining proposals are then assessed using criteria listed in the RFP for evaluation of the technical design, aesthetic and architectural appearance, qualifications, price, and agreements response (Table 9.3). For each of the above categories, a score from '1' to '10' points is assigned to each proposal. The best proposal in each category is assigned the top score of '10' points for that category. Each of the other three proposals is awarded a score based upon its comparison to the best proposal within each category.

Following the assignment of points in each of the above categories for each proposal, the score of each of the categories is multiplied by a weighting factor that reflects the local agency's decision of relative importance of the categories. For example, a weighting factor of '1' could be applied to the aesthetic and architectural appearance category. A weighting factor of '2' could be applied to the technical, qualification, and agreements response evaluation categories. A weighting factor of '3' could then be applied to the economic evaluation category, reflecting its perceived importance over the other categories. The total score is then determined, in our example, for each of the four proposals by adding together the total points times the weighting factor in each of the five categories. The maximum potential score is 100 points.

9.6 Negotiations process

Based upon the evaluation of the proposals and the ranking process prescribed in the RFP, the local government implementation team should seek authorization from its decision-makers to negotiate final proposed Design and Construction and Operations and Management Agreements with the top-ranked proposer. As an alternative, staff

Table 9.3 Illustrative criteria used to evaluate WTE proposals

Technical evaluation
• Feasibility and operational reliability of equipment and unit processes • Soundness of plan for project integration of processes and equipment • Contingency capabilities of proposed system • Demonstration of ability to comply and maintain compliance with all environmental regulations and permit conditions • Project expansion capability and ease of expansion • Safety design features and plan • Quality of residue produced • Adaptability of system to technological and regulatory changes • Management plan and construction schedule • Operating and maintenance plan
Qualifications evaluation
• Construction experience and applicability of experience cited • Operation management experience • Financial strength • Credit reports
Aesthetic and architectural evaluation
• Completeness of information • Architectural requirements in RFP • Overall visual appearance • Landscaping and site plan
Economic evaluation
• Net present value of disposal costs
Agreement response evaluation
• Level of acceptability of construction and operations agreements • Level of risk assumed by proposer • Time frame likely to be required to negotiate • Proposer's compliance with agreement conditions required to meet financing plan

Source: Reference [7].

could seek authorization to begin simultaneous negotiations with all of the ranked vendors with a time limit set for the maximum allowable negotiation period. If the negotiations are not completed within this period (which could be terminated at any time), the local government has the option of instructing staff to extend the negotiation period or enter into negotiations with the next ranked proposer.

Competitive negotiations between a community and a proposer for a WTE project requires considerable preplanning on the part of both parties. Each side needs to establish its negotiation objectives, its tactics, and counter proposals to the critical issues. Unfortunately, there is no formal procedure for undertaking negotiations for WTE projects. This does not mean that negotiations need be time-consuming. With proper pre-planning, the parties can strive to determine the critical areas of disagreement and their relative importance, and develop bargaining strategies for narrowing these differences.

There are three possible outcomes of any negotiation:

- Win/lose – One side is elated and the other side is a loser determined to get even;
- Lose/lose – Both sides are worse off than before and leave the table with distrust, frustration and hostility; and
- Win/win – Both sides are satisfied and get something.

The 'win/win' outcome is the most preferable in negotiations for the procurement of a WTE facility. Both the community and the proposer leave the negotiating table believing that they were successful in achieving their goals and objectives prepared at the outset. Agreements entered into under such conditions are much more likely to be respected by either party than under the 'win/lose' or 'lose/lose' outcomes.

References

[1] Feldstein Sylvan. Resource Recovery Revenue Bonds: An Analyst's Primer. New York: Merrill Lynch, Pierce, Fenner and Smith, Municipal Research Department; 1983.

[2] Yaffe Harold J, Wooten Jonathan. The development and financing of the Northeast Massachusetts (NESWC) Resource Recovery Project: a tale of twenty-two cities and towns. In: Proceedings of the 1984 National Waste Processing Conference. New York: American Society of Mechanical Engineers; 1984. pp. 102–10.

[3] Hayden John A. The full-service approach to resource recovery. Public Works 1983:68–70.

[4] Anonymous. Requests for proposals in State Government procurement. University of Pennsylvania Law Review 1981;130:179–215.

[5] State of New York. Resource Recovery Procurement: A Guidebook for Community Action. Albany: New York State Department of Environmental Regulation; 1980.

[6] Schoenhofer Robert F, Gagliardo Michael A, Gershman Harvey W. Fast track implementation of the Southwest Resource Recovery Facility. In: Proceedings of 1982 National Waste Processing Conference. New York: American Society of Mechanical Engineers; 1982. pp. 339–50.

[7] Hillsborough County, Florida. Request-for-Proposals for a Solid Waste Energy Recovery Facility. Tampa: Hillsborough County, Florida; 1984. pp. 6–1 and 6–5.

10 Ownership and financing of WTE facilities

Chapter Outline

10.1 Introduction

A discussion of WTE financing must first include a review of project ownership options. In general, the selection of the source of capital to fund the project, and the decision as to whether a government entity or private sector party owns the WTE facility, are usually decided together. This chapter will discuss the key factors involved in the selection of an ownership and financing plan for WTE projects. The ownership of a WTE project is one of the most important policy decisions that a community must make. Table 10.1 lists some of the major factors that must be considered in the choice of public or private ownership.

Waste-to-Energy. DOI: 10.1016/B978-1-4377-7871-7.10010-3

Table 10.1 Major decision factors affecting facility ownership

	Impacts under different ownership options	
Decision factors	**Public**	**Private**
Control over project	Yes	No
Operating risks	Low	Lower
Energy and materials revenues	Public owner	Private owner
Cost to system ratepayers: initial	Higher than private option	Lower than public option
Final	Lower than private option	Higher than public option
Residual value entitlement	Yes	No
Property tax payments	No	Yes

10.2 Ownership alternatives

10.2.1 Public ownership

In the United States, traditional government-owned, government-operated public services, such as schools, roads, water and wastewater plants, and similar facilities, are found in most municipalities or counties. Since this is the traditional approach to financing public works projects, all the legal and institutional mechanisms are usually in place when implementing a WTE facility in this manner.

Under this ownership alternative, the government entity bears all the operational risks, and expects, in return, to provide a waste disposal service at the lowest operating cost to its ratepayers. However, since local government is not a federal taxpayer, it is unable to realize the available tax savings from such normal business deductions as depreciation or investment tax credits if allowed under the federal tax code. The inability to use these tax benefits commonly results in a higher capital financing cost for publicly owned WTE projects. On the other hand, government would acquire a balance sheet asset that it would retain after the repayment of borrowed capital.

10.2.2 Private ownership

Private industry can play a major role in the development of WTE projects. The reason local government may consider private ownership is that the infusion of equity capital will reduce the project capital needs either through the financial impact of the initial private equity contribution, and thus, smaller bond size, or through annual equity contributions during the early years of the project. This is accomplished through project financing combining tax-exempt bonds, and possibly taxable bonds, and private equity, where private owners or third-party investors will obtain federal tax benefits associated with ownership of an industrial facility in addition to the attractive financial return. However, equity capital contributions must be structured in order for the private party to be recognized as the facility owners, and thus be eligible for any tax benefits. Local governments, faced with debt and budget limits and unwilling to assume technological risks, have increasingly transferred responsibility to the private sector.

Many WTE projects implemented in the United States are privately owned, although the availability of private equity capital in future years will be influenced by recent changes in federal tax and leasing laws limiting the tax benefits once available to non-governmental owners of WTE facilities. Consequently, private equity capital for WTE projects from such sources may be somewhat limited in future years.

Perhaps one of the most limiting factors in the development of privately owned WTE projects may be the availability of tax-exempt, industrial development bonds to finance the construction of such facilities. Since privately owned WTE facilities would have to compete with many other private uses of such bonds, there is a distinct possibility that private WTE project developers in some states may be unable to secure a bond allocation for their projects. On the other hand, government owned facilities would be exempt from this cap.

The major disadvantage of a privately owned facility is usually perceived as the fact that the local government will never own the WTE facility even though its taxpayers, through payment of the disposal fees, helped retire most of the bonds that paid for the construction of the facility. Furthermore, the municipality will have limited control over the facility, except to require periodic tests to demonstrate plant performance guarantees. Consequently, a privately owned, privately operated facility must be analyzed as a long-term provision of a service, not as a purchase of the project.

10.3 Prerequisite to financing

Prior to seeking financing for the WTE project a few requirements must typically be met:

- A ‘Put or Pay Waste Disposal Agreement’ must be secured for a large fraction of the project capacity. This type of agreement requires the project owner to give a guarantee that the waste flow will be available for the facility operator. Or the owner must be responsible to make the facility operator whole by paying the energy revenues that would be generated from the municipal solid waste;
- A Power Purchase Agreement (PPA), which must be secured and deposited and the start-up date carefully reviewed;
- A project site;
- Preliminary permit approval;
- Access to reputable grate technology;
- An EPC contract;
- An O&M operator agreement.

These above agreements will be needed in order to obtain access to cost-effective financing.

10.4 Financing options

WTE facilities are capital-intensive projects. Expenditures for such projects represent a significant and complex funding problem for any community, requiring a thorough evaluation of several alternative methods of financing. Most communities do not have the available capital out of their general revenue fund to 'pay-as-you-go' during construction of the facility. As discussed in the paragraphs below, the capital required to build a WTE facility may be raised from public sources, private sources, or combinations of public and private sources. Borrowing from such sources results in financing expenses related to the market interest rate on the bonds, or rate of return expected by private equity investors.

10.4.1 General obligation (GO) bonds

GO bonds are secured by the government's pledge of its full faith, credit, and taxing power. Although such bonds are secured by a pledge of *ad valorem* (property) taxes, they may be repaid with project revenues or any unrestricted income of the government entity, such as sales taxes, license fees, income taxes, and other fees. In most areas of the country, the use of a GO debt is contingent upon approval of local voters in a special referendum. Local governments have used GO bonds to finance such services as bridges, sewers, airports, stadiums, and housing projects.

GO bonds are generally considered the most secure form of municipal tax-exempt bonds, usually reflected in their relatively low interest cost. Additionally, a GO bond issue is the simplest and most readily marketable form of debt to issue since many of the complexities and potential delays associated with alternative methods of financing are not associated with GO bonds. One disadvantage, however, is that the government's borrowing capacity for other critical community projects may be restricted if such a pledge is undertaken for its WTE project.

10.4.2 Project revenue bonds

Revenue bonds are limited obligations of a community, only secured by the revenues expected to be generated from the operation of the facility. This includes all solid waste user fees and revenues from the sale of power and recovered materials, as well as investment income. Typically, the issuing government pledges in a bond rate covenant to fix and collect rates and charges for services rendered by the WTE facility sufficient at all times to pay operating expenses, bond principal, and interest. Such bonds are secured by trust indentures that control the use of the facility and sources of its revenues.

Generally, revenue bonds can be scheduled with maturities ranging up to 30 years. Revenue derived from the operation of the facility may be the sole security for the bonds. More commonly, additional security in the form of revenue from other sources must be pledged. In order to issue revenue bonds, the government's political body must adopt a bond resolution specifying the application of bond proceeds to construction of the facility, creating a lien on revenues of the facility, establishing a flow of funds and, in general, setting forth in detail the rights of the bondholders and obligations of the issuing government.

The rate of interest charged on such bonds, although higher than for GO bonds, is thus based upon potential investors' or bond rating agency (i.e., Standard and Poor's and Moody's) perceptions of the financial viability of the project; the contractual arrangements among the contracting parties; and perceived value of the local government's back-up pledge of revenues. Such perceived risks require that all project revenue bond financing must have debt service reserve accounts to protect against unforeseen revenue shortfalls. For example, the local government may be required to maintain a debt service coverage ratio of 1.0 to 1.5.

Industrial development revenue bonds (IDBs), a distinct form of revenue bonds, have been the most widely used form of financing for WTE facilities. These bonds are issued by local government to finance the construction of such plants which are then leased or sold to a private corporation.

Federal tax law, which is exceedingly complex and technical in nature, affects the issuance of IDBs to finance the construction of WTE facilities. The Tax Reform Act of 1986 has dramatically altered the use of IDBs to finance such projects. Following is a brief discussion of this tax area. The reader is cautioned that the community's tax counsel should be consulted for specific and detailed recommendations.

With respect to WTE projects, bonds issued by communities to finance such projects are generally regarded to be IDBs if one of the following is true about a project:

- Energy (steam, electricity, or recovered materials) is sold to a taxable entity such as an investor-owned utility or private company;
- A private contractor operates the project with a contract length in excess of five years;
- A private contractor bears a 'risk of loss' for situations beyond its control;
- Solid waste is delivered by private haulers (not agents of the government) and their revenues account for more than 25 per cent of the total project revenues.

At the time of writing, IDBs for publicly owned WTE projects are tax-exempt if 95 per cent of the bond proceeds are used for the solid waste disposal portion of the project. Under the so-called '95/5 rule,' expenses for the construction or installation of equipment related to the sale of byproducts from the plant (steam, electricity, or recovered materials) are non-exempt. Therefore, these expenses, which are typically 10 to 15 per cent of the total project cost, must be financed with taxable debt.

10.4.3 Grant funds, loan guarantees, and entitlements

There are a variety of programs in the United States to finance the implementation and construction of WTE projects.

10.4.3.1 Department of interior grants

The United States Department of the Interior, Office of Insular Affairs (OIA), offers a variety of specific grant funding opportunities to support WTE infrastructure needs in United States territories.

10.4.3.2 Capital Improvement Project (CIP) funding

CIP grant funding (also known as Covenant grants) have historically addressed solid waste infrastructure needs such as implementation of the scalehouse/weighing system at a landfill, and the purchase of automated, side-loader collection vehicles in recent years. CIP funding for WTE is subject to annual appropriations and is specifically for capital improvements in any of the four US territories. A unique feature of these grants is that they may be used to meet the local matching requirement for capital improvement grants of other Federal agencies, subject to OIA's approval.

10.4.3.3 Operations and Maintenance Improvement Program (OMIP)

This fund is used by OIA to promote and develop insular institutions and capabilities that improve the O&M of island infrastructure. This is the only OIA program that has specific legislative authority to require a matching contribution from the insular government.

10.4.3.4 American Recovery and Reinvestment Act (ARRA)

The American Recovery and Reinvestment Act (ARRA) of 2009 provides a number of new incentive programs that are intended to encourage accelerated commercialization and deployment of renewable energy resources and other innovative energy technologies.

10.4.3.5 Investment Tax Credit (ITC)

Tax credits have been available since 2008 for WTE facilities in the United States and are applicable to private developers.

10.4.3.6 Section 45 Production Tax Credit

Under ARRA, Section 45 of the Internal Revenue Code was extended and expanded. Among other provisions, the legislation extended and expanded the scope of Section 45 of the Code, which provides a tax credit for the production of electricity from wind, closed-loop biomass, open-loop biomass, geothermal energy, solar energy, small irrigation power, municipal solid waste, and refined coal. The production tax credit (PTC) gives a taxpayer a federal tax credit of 2.1 US cents per kilowatt of electricity generated at a qualified facility in the first ten years of operation. This facility must be owned and operated by the taxpayer, but the electric power must be sold to an unrelated person. As such, the PTC applies to all facilities in the United States, including United States possessions, but does not cover governmental entities like the American Samoa Power Authority (ASPA).

ARRA extends the placed-in-service date for wind facilities through December 31, 2012 for purposes of qualification for the PTC tax credit (10-year credit of 2.1¢/kWh in 2008). For closed-loop biomass, open-loop biomass, geothermal, landfill gas, WTE, hydropower facilities, and marine renewable facilities, the corresponding extension is through to December 31, 2013.

10.4.3.7 Treasury grant

Prior to the passage of ARRA, the PTC was the primary federal tax incentive for utility-scale renewable energy projects. As the PTC is a tax credit, the entity claiming the credit must have tax liability to credit the PTC against. Most developers do not have sufficient taxable income to take advantage of the PTC. The result is that developers often finance projects through tax equity financing, where an entity with sufficient tax appetite contributes equity to a project in exchange for the right to, among other things, the PTCs generated by the project.

Current economic conditions have severely undermined the effectiveness of these tax credits. As a result, the Recovery Act allows taxpayers to receive cash assistance from the Treasury Department in lieu of tax credits. As described by the Department of the Treasury, this funding will operate like the current-law investment tax credit. The Treasury Department assistance will be equal to 30 per cent of the cost of the renewable energy facility (the percentage depends on the type of facility) within 60 days of the facility being placed in service or 60 days after receiving an eligible application. The beneficiaries of the program are:

1. The project owners/lessees that receive funds;
2. Individuals and companies that are employed in the construction, O&M of projects; and
3. users of renewable energy.

To qualify for the Treasury grant under ARRA, the project must have been placed in service in 2009 or 2010. A special rule exemption extends this eligibility period for WTE facilities if construction started in 2010 and the facility is placed in service before 2014.

10.5 Private equity

Equity capital contributions from the facility developer or third parties, such as banks, insurance companies, corporations, and private investors, can reduce the overall capital needs for a particular project. For example, in return for an equity investment of some 15 to 25 per cent, the private equity contributor becomes the owner of the facility, and is entitled to tax benefits, if any, under existing law, and any accrued revenues from the project (e.g., tipping fees, energy and materials recovery revenues). This private owner also becomes the ‘deep pocket’ for repayment of the project bonds. A great many WTE projects have been financed utilizing contributions of equity capital from private investors. Due to the loss of favorable tax benefits to private investors under the Tax Reform Act of 1986, this level of private equity contributions in the WTE projects may be reduced in the years ahead. Additionally, the 10 per cent investment tax credit has now been eliminated, and depreciation for plant equipment has been extended from five to ten years. For those projects whose tax benefits were grandfathered under the liberal transition rules under this Act, equity contributions may continue to be available from private investors. The level of such equity contributions will depend in part upon the value of the tax benefits associated with the equity investment and the relative rate of return of these equity investments in WTE projects as compared to alternative investments.

10.6 Costs and facility operation

Typically, most WTE facilities run continuously (24 hours a day, seven days a week) with limited scheduled maintenance downtime for major components and systems. Intermittent operation or 'cycling' the incineration units up or down requires significant additional inputs of fossil fuels (propane, diesel fuel, natural gas) to heat up the incineration unit and air emissions control systems to proper operating temperatures. Heating and cooling oftentimes causes damage to these units, increasing repairs, and ultimately having an impact on operating and maintenance costs. An annual two-week shutdown for annual maintenance and overhaul is also typical for the combustion units as well as a periodic shutdown of the generator for maintenance.

10.7 Initial capital equipment

A typical WTE facility would require the following major components:

- Scale and scale house;
- An enclosed metal building and foundations with MSW storage area sized for two days' storage, control room, restrooms, break/lunch and meeting room(s), office, storage, and maintenance shop;
- Modular waste combustion units (primary and secondary);
- Air emission control systems (De-NO_X, selective non-catalytic (SNCR using dry urea for NO_X reduction), dry-absorbent injection system with sodium bicarbonate for acid gas neutralization, activated carbon injection to remove dioxins, furans, and heavy metals, and a filter bag house to collect scrubber consumables and any residual fly ash;
- Steam generating equipment (boiler), condensers, water cooling plant for condenser cooling, boiler feed pumps, multi-stage condensing turbine, electrical generator, electrical substation, and power transmission lines;
- Gearboxes;
- Water treatment plant for boiler makeup water;
- Ducting and ID fans;
- Ash handling system;
- Continuous emission monitoring;
- Master SCADA control system with back up computer;
- Data logging computer with capability for remote monitoring;
- Stack (50 foot) with height dependent on air permitting requirements;
- Site roadways, signs, landscaping, parking, and stormwater control system;
- Utilities, wastewater, potable water, stormwater, electricity, and telephone;

- Ancillary rolling stock (front-end loader, pickup truck, roll-off containers for storing ash streams, etc.);
- Building construction equipment (e.g., cranes and other lifting equipment);
- Equipment purchase, marine delivery to American Samoa, and installation;
- Administrative/staff time during facility development (construction monitoring);
- Permitting (federal and local);
- Shipping from the fabrication site to the installation location;
- Technical assistance for commissioning and startup testing;
- Testing of equipment and air emission control equipment.

Table 10.2 provides estimated capital costs for a proposed WTE facility in a US territory in the South Pacific, including a 15 per cent contingency and the applicable Treasury grant of 30 per cent, which can be applied to the overall capital costs for the project.

Table 10.2 Estimated capital costs for proposed WTE facility in American Samoa

Cost Items	Costs (US$)
Land[1]	0
Building (10,000 sq. foot building) @ US$200/sq. foot	2,000,000
Combustion, ash handling, and air emission control equipment	7,000,000
Waste heat recovery and power generation equipment and SCADA/ data recording/control systems	3,200,000
Scale house/weighing system	200,000
Utilities and substation/tie in	200,000
Rolling stock	200,000
Engineering, installation, testing	400,000
Permitting and implementation[2]	300,000
Freight	400,000
Subtotal	**13,900,000**
Contingency (15%)	2,085,000
Total capital costs	**15,985,000**
Treasury grant (30% of total capital costs)	4,785,500
Total capital costs – minus treasury grant	**11,189,500**

[1]Proposed parcel is government owned.
[2]Federal CAA, FAA and local permitting and procurement assistance.
Source: SCS Engineers.

10.8 Operating costs

Operating costs for a WTE facility depend on a number of factors including the following:

- Labor (plant manager, scale attendant, two to three plant operators, a maintenance mechanic);
- Costs of chemicals (lime, activated carbon, caustic) needed for air emissions control;
- Equipment regular maintenance and major repair and replacement (MRR) fund;
- Site and building maintenance;
- Utilities (water, wastewater, electricity, telephone, and Internet);
- Periodic air emissions testing;
- Ash disposal;
- Emergency fund;
- Insurance;
- Project management.

10.9 Estimated annual debt service and annual operating costs

Table 10.3 provides an example of a summary of debt service and operating costs for the proposed WTE facility in the South Pacific on an annualized basis. This Pro Forma financial model assumes that the owner would contract with a private vendor to own and operate the WTE facility, to enable the project to take advantage of the current Treasury Grant Program for solid waste facilities under the American Recovery and Reinvestment Act. This will allow a private vendor to receive a 30 per cent grant from

Table 10.3 Estimated capital and operating costs for a proposed WTE in American Samoa

Operating costs	Expenditure (US$)
Operating expenses:	
Vendor processing fee	869,670
Owner administration	25,000
Pass-through costs	767,044
Total operating expenses	1,636,714
Revenues:	
Energy production	2,225,196
Ferrous recovery	0
Non-ferrous recovery	0
Subtotal	2,225,196
Capital costs:	
From Table 10.2	11,189,500
Debt service:	1,314,314
Estimated tipping fee	25.04

Source: SCS Engineers.

the Department of the Treasury. As shown, this grant program would reduce the estimated capital cost of the facility (US$15.9 million) by nearly US$4.8 million. The model assumes that the private vendor would finance the cost of the remaining capital costs through issuance of private activity bonds with the help of the American Samoan Government or through a private bank/equity financing. For modeling purposes, we have assumed an interest rate of approximately 10 per cent. As shown, this would require an annual debt service payment of approximately US$1,314,315. Similarly to other WTE projects, it is further assumed that the owner would enter into a long-term contract for the purchase of the approximately 2.0 MWe of electrical power, which would be dispatched to the local power grid, as well as a tipping fee for all solid wastes, waste oils, and tires delivered by the local government to the facility.

There are potentially many ways that the project could be structured to enable the private vendor to receive an adequate return on private invested capital. This could include a high tipping fee and/or power purchase rate. For initial financial modeling purposes, we have computed the required initial power purchase price (US$0.152 per kWh) that the private vendor would have to receive from the local government and tipping fee (US$26.04), which approximates the current cost for the local government's landfill customers. The revenue stream generated by these two sources enables the private vendor to achieve his projected facility operating fees, typical pass-through costs for air emission control systems and utilities, and debt service, as well as providing an adequate return on his investment in the project. As shown, the owner would receive energy production savings of nearly US$0.15 per kWh compared to current costs, or roughly US$2.5 million annually.

10.10 Equipment life and replacement

The useful life of these WTE facilities is a minimum of 20 years, assuming proper maintenance, repair, and best solid waste industry operating practices.

10.11 Zero tip fee for a developing nation

In developing country, poor waste disposal practices in unregulated landfills often threaten potable water supply and health. These poor disposal practices are implemented with limited means at near zero tip fee disposal cost. To be viable and sustainable in these countries, TWE facilities will not be able to collect revenue from the waste stream and will have to rely solely on the revenue of energy sales and carbon credits.

If applying this 'US$0 tip fee' goal to US and European projects, electric power will need to be sold at a price between US$0.25 and US$0.35/kWh to support the project financial requirements. In developing countries, because the cost of labor is lower, there are opportunities to lower that figure. By applying some of the following factors, it is further possible to reduce the power price requirement without compromising the environmental compliance standards:

- Lower labor cost;
- Using local EPC contractor;
- Large project size;
- Adapting plant civil work to local needs;
- Improving waste heating value;
- Improving energy recovery efficiencies;
- Favorable financing terms.

In most favorable cases, a 'US$0 tip fee' WTE project could be made viable with power sales as low as US$0.12 to US$0.14/kWh.

11 O&M of WTE facilities

Chapter Outline

11.1 Introduction

The O&M of WTE facilities requires special consideration and operator skills. WTE facilities are power plants that require combined expertise in the combustion of solid fuel and in handling of corrosive flue gas. The O&M of WTE has to meet stringent power industry environmental regulations on waste, air, and wastewater. The WTE industry is now well known and it is possible to implement a comprehensive operation plan that carefully anticipates risks and pitfalls specific to the WTE operation. A few private and specialized full-service operation providers are also available to ensure long-term operation success of the plants.

11.2 Key aspects of the O&M approach for WTE facilities

11.2.1 Selecting the operator

A few decision criteria are assessed when selecting the operator, including: operator expertise and references, O&M contractual term, expected operation guarantees, scope of services, and cost.

The responsibility of operation can be carried by the owner, private subcontractor, or the EPC contractor. A long-term operation contract from an experienced operator offers the best guarantee for the plant owner. A short-term contract will shift

Waste-to-Energy. DOI: 10.1016/B978-1-4377-7871-7.10011-5

maintenance risks to the owner. In either case, success of the operation will rely on the quality of the facility staff, who should be recruited for their expertise and solid long-term approach, thus avoiding WTE operation pitfalls.

The O&M contract defines responsibilities between the owner and operator. Typically, changes-in-law risks stay with the owner, but environmental compliance is mainly the responsibility of the operator who shares liability with the owner. Operator minimum performance criteria is defined. A variable level of bonus payment to the operator is advisable to align the operator's interest with the plant owner's. The variable payment should be linked to tonnage throughput and amount of energy recovered. A reputable operator will be able to provide the performance guarantees that will support project-financing requirements.

Typically, performance bonds or hold-back of periodic payments are incorporated, to insure against the operator maintaining their contractual agreement or their becoming insolvent. Operators are required to abide by local laws and regulatory authorities in accordance with the country and locale of the plant. Insurance coverage is required by the contract, with the owner to cover any damage or loss to the equipment and structure. Insurance for 'Business Interruption' remains solely with the operator, as does theft. Special consideration should be given to planning the major repair and replacement (MRR) of plant and equipment with the allocation of a sustainable budget.

11.2.2 Commissioning

Most successful WTE facility operations can be credited to the quality and success of the commissioning phase where the long-term operator was involved during the design and construction of the facility. This is highly recommended. The operator should validate general design, operation, and maintenance access of the equipment and commissioning plan. Preparation of the commissioning should begin at least 12 months prior to the plant start-up, where the operator prepares the operation plan that is specific to the project, including staffing requirements. Each operator position should be defined with job description, job qualifications, and physical requirement. Safety plan and environmental procedures should be established. The operator should review and confirm operation plan and procedures provided by the EPC contractor.

At least three months prior to the commissioning, operators and key maintenance staff shall be recruited and formal operator training shall begin. The commissioning is typically performed under the supervision and responsibility of the EPC contractor.

Commissioning a WTE facility can be described as requiring two different phases of implementation.

11.2.2.1 Phase 1

The first phase of commission requires physical checks of Equipment and Systems. This phase becomes quite complex as the facility is made up of a lot of equipment from various suppliers, each with their own installation and operation technique and warranty, many through sub-contract installations. A Commissioning Engineer is assigned to co-ordinate all of the post installation activities to ensure that the integrity of any warranties is maintained and ultimate performance ensured. Oil levels, lubrication, correct rotation of equipment, packing and fills, control loops and terminations, alarm levels and so on are systematically checked, corrected and signed-off by the Engineer in advance of the plant being started. This role will work in co-ordination with the equipment supplier's own Sales Technician or Engineer.

11.2.2.2 Phase 2

The second phase of commissioning requires Performance Testing of the entire facility to ensure that the obligations of design, construction, and performance are met. These requirements form a portion of the original Design and Build Contract. This phase cannot be attempted until the completion of the former and will require the operator to operate the plant. This phase must be very carefully completed under the experienced eye of both the commissioning engineer and the operator. Typically, the operator would be hired to conduct the performance testing by the designer/builder and under their supervision. This then forgoes the potential liabilities between the two parties should problems arise from operational mishap.

Performance testing protocol shall be agreed with the EPC contractor in accordance with its contractual obligations. Testing should be long enough to verify plant reliability and stable operation. In some case, performance testing lasts for a few months, with tracking of 'defects' lasting for 12 months. Tracking the defects forms a prioritized list of items that require some negotiation to correct. This list may be comprised of items that were 'missed' by design or installation, underperformance of equipment or sub-systems, equipment failure, poor installation of equipment, or equipment supplied 'not as ordered.'

It can be a timely affair to rectify some of these situations, many of which can and will delay the start-up of the facility to regular operation. Contractual safeguards are required to prevent the operator from being penalized for, or from, this delay; the penalty is the cost of an idle workforce. These safeguards usually involve the Design/Build contractor being responsible for the operator's payroll and, potentially, penalties from the owner due to the late start. This of course depends on the severity of the situation.

11.2.3 Plant performance, optimization of economics

WTE facilities are capital-intensive and will need to be operated at maximum capacity every day to keep economics at a reasonable level. Good performance not only secures solid plant revenue but, to some extent, high performing plant also costs less to run. Unscheduled downtime and reactive maintenance are very costly.

WTE performance is driven by the following five performance factors:

1. *Boiler availability.* Boiler availability is measured by the number of hours that the boiler is operated above 50 per cent of maximum continuous rating (MCR) divided by the number of calendar hours for the period. As an average, the industry operates at 87 per cent availability with a solid performer running at 90 per cent. A few excellent plants are able to run at

around 92 per cent for a few years. To keep unscheduled downtime under control, the units should be inspected once or twice a year and tubes, refractory, and equipment need to be replaced on time during carefully planned scheduled outages.

2. *Boiler mass load.* This is the boiler throughput in comparison to design MCR load. Mass throughput is typically limited by a waste heating value that is higher than designed, grate burn-out that is above 3 or 5 per cent, or air/flue gas flows which are caused by limited fan capacity or boiler fouling.
3. *Boiler heat load.* This is the boiler energy output that is typically measured in steam flow in comparison to design MCR load. The maximum heat load is determined by the design and manufacture of the grate system. The heat load is typically limited by a waste heating value that is lower than designed, combustion stability or air/flue gas flows that is caused by limited fan capacity or boiler fouling.
4. *Steam cycle efficiency.* This is the efficiency of converting steam into electricity or a heat source. Excessive in-house stream consumption and low turbine efficiency can limit the overall energy output.
5. *In-house parasitic load.* This is the internal electrical load needed to power plant equipment. This internal load is subtracted from the turbine gross output. Only the net output (gross output minus internal load) can be sold to the grid and generate revenue.

The project will have to operate within *all* designed performance factors to be able to provide projected revenues during the project-financing phase.

11.2.4 Operation risks and remediating them

The risks of operating a WTE facility can be classified in three areas:

1. *Health and safety.* Many issues of health and safety start at the design stage: access to equipment by forklifts, access and egress throughout the plant for personnel, chemical and oil storage, fire protection, and safety showers, to name a few. Local regulations and laws may dictate what the basic requirement is. For example, waste bunker fire protection requires special consideration, including installation of fire cannons or water deluge systems. Due to the moving equipment and high temperature surfaces, the safety of plant personnel must be well-organized with a program that must include solid lock out/tag out of equipment, and confined space procedures.
2. *Environmental risks.* Because WTE plant operation is regulated by complex waste, air, and water regulations, most facilities have an environmental engineer on staff to ensure understanding of permits, establishment of operation procedure, training of personnel, emission testing, and reporting to authority. The environmental systems will need to be thorough. ISO 14001 is a tool that can be used to ensure sustainable compliance; it is a voluntary international standard that establishes the requirements of an environmental management system. Environmental regulations can be expected to become more stringent through the length of a long-term operating contract. The operator must be prepared to seek out and understand new technologies and testing methods. Many owners in the public sector have direct access to monitoring emissions and corresponding excursions in those limits, remotely from the plant. This puts the operator in a completely supervised situation.
3. *Maintenance and equipment breakdown.* A modern WTE facility today can expect to have a 25-year lifetime before a major rebuild to the grate, furnace, boiler pressure parts, or turbine/generator. Most build, operate, and transfer (BOT) contracts reflect this in their duration. However, due to the corrosive nature of the WTE boiler flue gas with dew point

temperatures very close to the operating range, special considerations are required to prevent frequent breakdowns. Boiler failure or flue gas treatment corrosion can be expensive to correct if left unattended. Boiler tube corrosion can be mitigated by improving combustion, controlling flue gas temperature, or by protecting boiler tubing with a high heat transfer refractory or high nickel/chrome overlay materials. Flue gas treatment equipment needs to run within design temperature and flow to avoid erosion or Dew point corrosion. As in any typical power plant, failure of turbine, generator, or transformers is not only expensive, but leads to extended plant outages.

Fugitive dust from the air pollution control (APC) plant is not only a health hazard, but has long-term effects due to corrosion on parts of the plant not normally subjected to corrosion, structural parts of building and boiler supports included. This 'dust' is actually a reacted component of the flue gas scrubbing system and is high in levels of chloride and sulfur compounds.

Two types of maintenance systems or programs are employed in WTE facilities:

1. *Preventative maintenance (PM) program.* A PM program consists of a scheduled series of equipment checks carried out either under operation or during a scheduled shutdown. These checks can consist of many items, some quite simple, others a little more complex. Some of these items, for example, would be lighting inspection, equipment lubrication, filter replacement, crane wire ropes and electrical cables, tube thickness checks, air heater cleaning, flue gas ductwork, and scrubber nozzles and atomizers, to name a very few. This program is housed in software that enables the checks to be carried out on a scheduled basis with a record of history and replacement. A good PM program will identify the requirements if and when major maintenance is required.
2. *Major maintenance (MM) program.* An MM program consists of items that may or may not be re-occurring, but certainly with re-occurrence longer than one year. A five-year MM program schedule is normally modified and produced annually. Items that would be included in this schedule would be large motor replacement/rewinds, turbine overhauls, boiler pressure parts replacement, air compressor rebuild, boiler feed water pump rebuilds, and so on. Through proper PM programs, the MM program can be monitored and modified each year through the budgetary process to improve plant performance and return. The MM program is a significant portion of the overall maintenance expense of the plant. Significant adjustment to O&M statements can be made as to how these costs are either depreciated or expensed.

11.2.5 Approach to combustion control

Steam flow has been, and still is, used in power plants to control the fuel flow when fuel is very homogeneous, which gives very predictable performance. This very predictable performance provides an extremely stable operation, and the fuel is usually metered accurately with a mass flow meter when fuel is liquid, a gravimetric feeder in the case of coal, or an ASME flow nozzle for gas fuels. Because of the consistent heat value of the fuel, the steam flow could be used to set the fuel flow – or even the plant power output could control the fuel flow. The total air flow could then be set for the lowest boiler outlet O_2 compatible with the NO_X and the CO level to achieve. Because of the stability of the fuel composition, the boiler outlet O_2 could be trimmed down to less than 3 per cent on solid fuels like coal and down to 1 per cent on gas fuels like natural gas and the plant operated within a very narrow O_2 range. This stability allows these plants to work in dispatch mode, with the power demand from the utility dispatcher literally controlling the fuel feed. These plants can be modulated and follow the load demand fluctuations.

This approach, however, does not work well in WTE facilities with a fuel which has a highly variable composition and heat content that cannot be metered accurately. As a result, the steam flow constantly changes because the fuel characteristics change. The lag between the fuel feed increase or decrease and the steam flow is very big and very often out of phase. The steam flow may be calling for a decrease in fuel feed because of flare-up from a batch of fuel with a very high Btu content, while the next batch of fuel being fed could be very low in Btu content. The steam flow will cause the decrease with some lag. This will cause a big dip in the steam flow resulting from the decrease of the fuel flow combined with the low Btu content. In the meantime, the flare-up causes the flue gas temperature to rise and the excess O_2 to drop and possibly CO is generated. The over fire air and the excess O_2 control will try to correct the situation and likely get in conflict. This system has three variables (steam, fuel feed, and secondary air flow) that interact with each other with long lags. The result is that the system is unstable, constantly swinging up and down on load, with steam flow going well above and well below the set point, resulting in high furnace temperatures, and high CO or NO_x in the flue gas. This situation can be mitigated somewhat if the fuel is very well blended, thereby having a consistent heating value.

An 'O_2 combustion control' system eliminates the instability described above. The steam flow measurement has no control applications. The steam flow is the result of the complete combustion of the fuel on the grate. The operator selects the amount of steam generation needed, usually the maximum unit capability. Waste-to-energy plants always run at maximum steam flow. This selection provides a calculated primary air flow set point. This set point is an input into the primary air fan damper controller that will adjust the damper to maintain that air flow.

The secondary air flow set point is similarly determined and maintained. The ratio of secondary air to total air is predetermined from tests to correct for CO or for seasonal fuel quality variations. The secondary air is further divided into that for front and rear nozzles; again this ratio is determined from tests run during commissioning of the plant. The total air flow does not vary and the measured steam flow has no control applications.

The fuel feed is controlled with the excess O_2 in the flue gas coming out of the furnace. At least two O_2 analyzers of high quality are recommended. The oxygen analyzers have an extremely rapid response and the lag between the O_2 changes as a function of the fuel feed quantity and quality are very small. The only variable is the fuel and there is no interference from the steam flow measurement process. The fuel flow will be increased when the excess O_2 goes above the set point and will be decreased when it goes below it. If the O_2 is maintained at the set point value, it means that the right amount of combustibles (regardless of the volume or the weight of the garbage) is being fed in and burnt, and the resulting steam flow will be at the right value.

11.2.6 Long-term approach

The long-term sustainable operation of a WTE facility will be the result of a careful balance between performance, risk remediation, maintenance planning, and operator training. In plants that are owned by the public sector and operated by the private sector, long-term contracts (20–25 years) and independent plant audits provide a more stable platform for all parties. It is not uncommon for private operators to 'invest' in the plant for better returns for the long term.

On-going training of operators should be implemented in the area of safety, environmental compliance, and boiler/equipment operation. Changes in government regulations, whether environmental or health/safety, require updating of training programs. Certification to national, state, and provincial standards for the operation of these types of plant and steam boiler is imperative. Management responsibilities for training must also provide for progression, expansion, and technological change.

Maintenance planning should include a solid planned outage program to carry out inspection and replacement with a preventive maintenance program in place. A reasonable budget needs to be established for the long term. Safety and environmental issues must be treated with the highest priority. Plant staff know-how with multi years experience in the WTE industry is necessary to identify risk, and bring best practices to the project.

Appendix A
WTE Case Studies

Hillsborough County, Florida

Rated combustion capacity: 1,800 tons per day

Unit design: Three 400 ton per day and one 600 ton per day waterwall furnace

Energy generation at rated capacity: up to 29 MW, sold to Tampa Electric Company

Overview

The Hillsborough County Resource Recovery Facility began commercial operations in October 1987 after more than a decade of implementation planning. The facility processes 1,800 tons per day of municipal solid waste, while generating up to 47 MW of renewable energy. The facility is owned by Hillsborough County, which supplies all of the waste. The operator of the facility is Covanta Hillsborough, Inc. under long-term agreement with the County. The facility is located near Brandon about 10 miles east of the City of Tampa. This facility uses secondary sewer treatment effluent from an adjacent wastewater treatment plant for part of the process and ash cooling process water. Installation of new emissions control equipment to comply with Federal Clean Act amendments was completed on August 26, 2000.

County's solid waste system

The WTE plant is the cornerstone of the County's integrated solid waste management system, which includes curbside collection and recycling of such materials as plastic and glass containers, aluminum cans, and newspaper. The County delivers MSW from two solid waste transfer stations. Ash from the facility and bypass materials is delivered to a County-owned landfill.

Facility design

The Martin reverse reciprocating grate system is used for combustion of the incoming MSW. The boiler design includes 600 psig/750°F (400°C) superheater outlet conditions. The following air emission control equipment are utilized: semi-dry flue gas scrubbers injecting lime, fabric filter baghouses, nitrogen oxide control system, mercury control system, and continuous emissions monitoring (CEM) system.

Facility costs and finances

The initial 1,200 ton per day facility was constructed at a cost a little over US$80 million with proceeds from a system-wide bond issue. This system has consistently generated operating surpluses since the mid-1990s. Nearly half of the system's revenues are derived from special assessments on properties receiving residential collection and disposal services within the service area. Assessments are included on the property tax bills and constitute a lien, claim or charge held by one party, on property owned by a second party, as security for payment of some debt, obligation, or duty owed by that second party on property. Approximately 26 per cent of system revenues are derived from tipping fees remitted to the county by three franchise haulers serving commercial and industrial customers. Other revenue sources include sales of electricity to Tampa Electric.

City of Baltimore, Maryland

Rated combustion capacity: 2,100 tons per day
Unit design: Three 750 ton per day waterwall furnaces
Steam production: 510,000 pounds steam/hr @ 850 psig/825°F (440°C)

Overview

The Baltimore Refuse Energy Systems Company (BRESCO) has been providing disposal of up to 2,250 tons per day of municipal solid waste from Baltimore City, Baltimore County and other areas in Maryland since 1985. The Northeast Maryland Waste Disposal Authority had entered into an agreement with WESI Baltimore, Inc. and later BRESCO for the design, construction, ownership, and operation of the facility. Pursuant to this agreement BRESCO is entitled to sell up to all of the energy produced at the facility to the Baltimore Gas and Electric Company (BGE).

Facility design

Trash is delivered into an enclosed receiving pit where a clamshell crane is used to pick up three to four tons of waste at a time and feed the waste into one of three processing units. Each unit is made up of a furnace, a boiler and an air pollution control system. The reciprocating motion of the grates (Von Roll) inside the furnace moves the waste through the unit, insuring complete combustion. The primary combustion air is drawn from the refuse pit area, sustaining a negative pressure in the unit. The negative pressure prevents garbage odors and dust from escaping into the environment. Surrounding the grate systems are large utility-type boilers, which recover and 'recycle' thermal energy released during the combustion of the waste. This recycled energy is recovered in the form of high-pressure steam. At full capacity, the plant can generate in excess of 500,000 pounds of steam per hour. Part of the steam is used to make electricity and the rest is used for district heating and cooling. BRESCO is capable of supplying up to 300,000 pounds of steam per hour to Trigen, which distributes the steam to buildings in downtown Baltimore. BRESCO can also produce up to 60,000 kilowatts per hour for sale to BGE. The BRESCO WTE facility successfully reduces the volume by approximately 90 per cent. BRESCO also recovers ferrous and non-ferrous metals from the ash residue. These metals are shipped off-site to be recycled. The ash residue is approximately 28 per cent, by weight, of the incoming waste. The ash is used by the City of Baltimore for alternate daily cover at its Quarantine Road landfill.

Facility costs and finances

The facility was financed in 1982 with the proceeds of a revenue bond and private equity (US$254 million). An additional US$40.1 million in revenue bonds were approved in 1998 to fund the retrofit of the existing air pollution control systems to comply with the Federal Clean Air Act Amendments.

City of Commerce, California

Rated combustion capacity: 2,100 tons per day
Unit design: Three 750 ton per day waterwall furnaces
Steam production: 510,000 pounds steam/hr @ 850 psig/825°F (440°C)

Overview

Planning for the Commerce refuse-to-energy facility began in 1981. The original goal of the project was to demonstrate that WTE is a viable alternative method of solid waste management in California's South Coast Air Basin, where air pollution requirements are the toughest in the world. The facility has achieved its goal, having been in successful operation since 1987. The facility burns an average of 360 tons of trash per day and generates 10 MW (net) of electricity for sale to the Southern California Edison Company. This is enough electricity for 20,000 Southern California homes. The facility also provides certified destruction services for classified or sensitive documents and materials.

Facility design

The Commerce facility was the first plant in the world to use a unique state of the art combination of air pollution control devices. These devices consist of ammonia and limestone injection into the furnace, followed by a dry scrubber and finishing with a baghouse. This combination of devices has earned the Commerce plant the reputation of being among the cleanest of all the plants of this type in the world.

The Commerce refuse-to-energy facility has also established itself as one of the best all-round refuse-to-energy plants in the world, having produced some of the lowest emissions on record and operating an innovative ash reuse system. The system has won four national awards:

- Environmental Protection Award from 'Power' magazine;
- Award of Excellence from the Solid Waste Association of North America (SWANA);
- Grand Prize for Operation/Management from the American Academy of Environmental Engineers;
- Facility Recognition Award from the American Society of Mechanical Engineers.

Facility costs and finances

The facility was financed in 1982 with the proceeds of a revenue bond and private equity (US$254 million). An additional US$40.1 million in revenue bonds were approved in 1998 to fund the retrofit of the existing air pollution control systems to comply with the Federal Clean Air Act Amendments.

City of Spokane, Washington

Rated annual combustion capacity: 348,200 tons per year

Unit design: Two 400 ton per day waterwall furnaces; Von Roll reciprocating grates

Electrical production: 26 MW (gross); 22 MW (net) electricity

Overview

Environmental concerns associated with continued landfilling, new Washington State solid waste regulations, and federal actions directly affecting Spokane area landfills led the City and County to jointly develop a comprehensive program for regional solid waste reduction, recycling, recovery of energy, and residue disposal. This cooperative effort resulted in the Spokane Regional Solid Waste System, which includes the WTE facility, North County Transfer Station, and Valley Transfer Station. Ash produced from the WTE process is sent to an ash mono fill at Rabanci's Roosevelt Regional Landfill in Klickitat County, WA. In addition, there is an active cell at the Northside Landfill that is available for bypass and non-processible materials collected by the system. These facilities are intended to provide long-term, environmentally sound solid waste disposal for both the city of Spokane and the other incorporated and unincorporated areas of the County.

Facility design

The plant is designed to combust municipal solid waste using a Von Roll reciprocating grate system with two process lines (each designed to combust 400 tpd). The boilers use an overhead refuse crane and a ram feeder. Combustion temperature is designed at 2,500°F (1,370°C) using natural gas for startup. Steam flow to the turbine is 188,000 pounds per hour @ 830 psig/825°F (440°C). Ash handling utilizes a semi-dry, vibrating pan conveyor system with ferrous and non-ferrous metals recovery. Air quality control is accomplished using a dry scrubber, fabric filter, De-NO_X system for each boiler line.

Facility costs and finances

Permanent long-term financing was secured in January 1989, at which time the US$50 million short-term notes were paid off. The City of Spokane borrowed US$105,250,000 in revenue bonds to finance the cost of acquisition and construction of the waste to energy facility, two transfer stations, recycling centers, household hazardous waste turn-in sites, and a landfill cell for disposal of bypass and non-processible materials. The City of Spokane and Spokane County received US$20 million of the revenue bonds for landfill closure expenses.

In addition to the revenue bonds, the System was financed by a US$60 million Referendum 39 Grant from the Department of Ecology. The City Council approved acceptance of this grant on November 17, 1986, and the County Commissioners approved it on November 18, 1986. On November 24, 1986, the US$60 million grant, which provided 50 per cent matching funds for eligible expenditures, was executed.

Pinellas County, Florida

Rated combustion capacity: 3,150 tons per day
Unit design: Three 1,050 ton per day waterwall furnaces
Steam production: 510,000 pounds steam/hr @ 850 psig/825°F (440°C)

Overview

The Pinellas County facility began operating in May 1983, originally as a two-unit plant with a nominal processing capacity of 2,000 tons per day. A third combustion unit was brought online in 1986 increasing the nominal capacity to 3,150 tons per day, making it one of the largest WTE facilities in the world. The plant was operated for over 20 years by Wheelabrator Pinellas, Inc. under a consolidated management agreement. Under a procurement process, the County awarded a 17-year plant operating agreement to Veolia, Inc.

The facility processes about one million tons of garbage every year and can produce up to 75 MW per hour of electricity. About 60 MW was sold to Progress Energy for distribution within the community, and the remainder powers the plant itself. This electricity powers approximately 45,000 homes and businesses every day. Since its opening in 1983, the plant has combusted more than 22 million tons of MSW, enabling the county to extend the life of its one remaining landfill.

Facility design

Surrounding the grates (Martin reciprocating grates systems) are large utility-type boilers, which 'recover and recycle' thermal energy released during the combustion of the waste.

This recycled energy is recovered in the form of high-pressure steam. At full capacity, the plant can generate in excess of 500,000 pounds of steam per hour. Part of the steam is used to make electricity and the rest is used for district heating and cooling. Ash generated from the combustion of solid waste is transferred to the adjacent residue storage and processing building. Here, the ash is size-separated using screens, and both ferrous (steel) and non-ferrous (aluminum) metals are recovered from the ash using mechanical equipment such as magnets and eddy currents. The recovered metals are sold to steel mills and smelters for recycling, and the remaining ash is used for landfill cover and interior site berms and roadways.

Facility costs and finances

Facility construction was financed by a series of revenue bonds, which were defeased in 2005. The current system tipping fee is US$37.50 per ton.

City of Portsmouth, Virginia

Rated combustion capacity: 2,000 tons per day
Unit design: Three 750 ton per day waterwall furnaces
Steam production: 510,000 pounds steam/hr @ 850 psig/825°F (440°C)

Overview

The Refuse Derived Fuel (RDF) plant began operation in 1988 as part of the Southeastern Public Service Authority's (SPSA) WTE system. It processes nearly half of all solid waste received by SPSA. The RDF plant is designed to process 2,000 tons of waste a day.

Facility design

Waste delivered to the Portsmouth-based plant is unloaded on a 1.3-acre tipping floor, where it is pushed onto conveyors to begin processing. Employees on the tipping floor and cameras located in the control room screen the waste for hazardous materials or bulky items that could endanger or impede the processing of the waste.

Equipment is then used to screen, size, separate, and shred burnable waste into a uniform four-inch particle size. In addition to refuse derived fuel production, non-processible waste and reject materials are removed from the RDF plant waste stream, and ferrous metals and aluminum cans are separated for recycling. A system of belt magnets extracts over 1,000 tons of ferrous metals monthly for recycling.

The RDF is transported to an adjacent power plant to burn 1,500 tons of RDF daily. It produces all of the process and heating steam as well as the majority of the electrical power required by the Navy's largest shipyard, the Norfolk Naval Shipyard. Electrical power in excess of the shipyard's needs is sold to Virginia Power.

The power plant utilizes spray dryer absorbers and fabric filters have replaced electrostatic precipitators, allowing the facility to exceed the most stringent EPA requirements.

Facility costs and finances

Wheelabrator Technologies Inc., a wholly owned subsidiary of Waste Management, purchased the SPSA refuse derived fuel plant and adjacent WTE facility. The closing of the sale followed votes taken by the SPSA Board of Directors on 28 April 2010 as well as the transfer of US$150 million in funds from Wheelabrator to SPSA. After narrowing the candidates to two, SPSA chose Wheelabrator in November 2009. SPSA will use the US$150 million from the sale to pay down debt to the Virginia Resource Authority (VRA) and other lending institutions, and repay money owed to the City of Virginia Beach. As a result of the transaction, 164 SPSA employees at the two Portsmouth facilities will become Wheelabrator employees. Wheelabrator also plans to invest more than US$20 million in capital improvements.

RenWu, Taiwan

Rated combustion capacity: 1,350 tons per day
Unit design: Three 450 ton per day waterwall furnaces
Energy generation at rated capacity: up to 31 MW, sold to TaiPower

Overview

The RenWu facility began commercial operations in December 2000. The facility processes 1,350 tons per day of combined municipal solid waste and commercial waste, while generating up to 31 MW of renewable energy. The facility is owned by the local Government, which supplies all of the waste. The operator of the facility is Sita Waste Services, Ltd. under long-term agreement with the Government. The facility is located near Kaohsiung City, south of Taiwan.

Facility design

The Martin reverse reciprocating grate system is used for combustion of the incoming MSW. The boiler design includes 40 bars/750°F (400°C) superheater outlet conditions.

No effluent is generated from treatment of wastewater as wastewater is completely reused. The high-concentration refuse bunker leachate is injected directly to the incinerator for burning. Wastewater coming from vehicle washer, daily living, and recycling is reused after physical, chemical, and biological treatment. Hence, there is no concern for wastewater pollution.

The most stringent design for dioxin emission standards in the world, 0.1 ng TEQ/NM^3, and the best available control technology (BACT), are adopted for the plant's flue gas treatment system. This system includes a semi-dry scrubber, activated carbon injection device, SNCR, and bag filters. The treated flue gas is emitted into the atmosphere through the 120-meter stack. Therefore, the impact of treated flue gas on the air quality around the plant environment is minimal.

The RC building enclosing plant equipment effectively achieves acoustic and vibration control. Moreover, most equipment is installed with noise reduction device/design. The noise levels in the plant and along the site boundary are in complete compliance with national noise control criteria. The refuse bunker is maintained under a negative pressure condition to trap odor. The odor produced in the refuse bunker is drawn by fan into the incinerator as the primary air. The refuse bunker leachate is injected directly into the incinerator and burned. Thus, no odor is released.

Ash generated from waste incineration is transported directly to the designated sanitary landfill for disposal. Fly ash is solidified before delivery to the landfill for disposal.

Rozenburg, the Netherlands

Rated combustion capacity: 3,500 tons per day

Unit design: Six 450 ton per day waterwall furnaces and one 800 ton per day waterwall furnace

Energy generation at rated capacity: up to 57 MW, sold to the local power grid

Overview

The Rozenburg facility began commercial operations in December 2000. The facility processes 3,500 tons per day of combined municipal solid waste and commercial waste, while generating up to 57 MW of renewable energy. The facility is owned by Van Gansewinkel Groep, which supplies all of the waste. The operator of the facility is AVR, a subsidiary of Van Gansewinkel Groep. The facility is located near Rotterdam in the Netherlands.

The WTE plant in Rozenburg received an accelerated R1-status from the Minister of the Environment, Jacqueline Cramer, and has become an 'installation for useful application.' The award arrives ahead of the guidelines that will come into force at the end of 2010 across the EU. As a result, Van Gansewinkel Groep can receive waste more easily from abroad and incinerate it at the AVR facility in Rozenburg.

Facility design

The DBA six roller grate system is used for combustion of the incoming MSW. The boiler design includes 40/28 bar superheated steam outlet conditions.

The superheated steam drives three turbines. The turbines drive generators capable of producing up to between 28 MW and 57 MW. Around 30 MW of this power is used internally on the site. The excess is exported to the local grid. Steam is exported to nearby industries.

Rozenburg extensive air pollution control system includes: electrostatic precipitator filters, two-step wet scrubbers, a Coke filter, and De-NO_x SCR system to meet stringent European regulations.

Nice, France

Rated combustion capacity: 1,300 tons per day
Unit design: Four 325 ton per day waterwall furnaces
Energy generation at rated capacity: up to 28 MW, sold to French utility EDF

Overview

The Nice facility began commercial operations in 1979 with two units. The third and fourth units were added in 1982 and 1997, respectively. The facility processes 1,300 tons per day of combined municipal solid waste and commercial waste (5,000 tons per year of hospital waste are also treated and incinerated on site), while generating up to 28 MW of renewable energy and providing steam to an extensive district heating system. A sludge drying system was added in 1987 to facilitate the incineration of sludge within existing units. The facility is owned by the City of Nice, which supplies most of the waste. The operator of the facility is Sonitherm, a subsidiary of Veolia Environment, under long-term agreement with the government. The facility is located in the City of Nice, South of France.

Facility design

The Martin reverse reciprocating grate system is used for combustion of the incoming MSW. The boiler design includes 40 bars/400°C superheater outlet conditions providing steam to two turbines combining 14 MW output and to district heating. The air pollution control system was upgraded in 1998 to include electro-precipitators and wet scrubbers.

Sludge drying and incineration capacity is 100 tons per day. Two auxiliary boilers of combined 40 MW capacity provide full back-up energy supply to district heating during incineration unit outages.

The plant provides energy to a population of 11,000 through three district heating loops for a capacity of 95 MW.

Ferrous and non-ferrous metals are recovered on-site from the bottom ash.

Index

CPSIA information can be obtained at www.ICGtesting.com
Printed in the USA
LVOW071222270613

340437LV00005B/25/P